INTRODUCTION TO
Solid Modeling
Using SolidWorks® 2009

William E. Howard
East Carolina University

Joseph C. Musto
Milwaukee School of Engineering

 Higher Education

Boston   Burr Ridge, IL   Dubuque, IA   New York   San Francisco   St. Louis
Bangkok   Bogotá   Caracas   Kuala Lumpur   Lisbon   London   Madrid   Mexico City
Milan   Montreal   New Delhi   Santiago   Seoul   Singapore   Sydney   Taipei   Toronto

The *McGraw·Hill* Companies

 Higher Education

INTRODUCTION TO SOLID MODELING USING SOLIDWORKS®, 2009

This book is printed on recycled, acid-free paper containing 10% postconsumer waste.

1 2 3 4 5 6 7 8 9 0 QPD/QPD 0 9

ISBN 978-0-07-337540-3
MHID 0-07-337540-3

Global Publisher: *Raghothaman Srinivasan*
Senior Sponsoring Editor: *Bill Stenquist*
Director of Development: *Kristine Tibbetts*
Developmental Editor: *Lora Neyens*
Senior Marketing Manager: *Curt Reynolds*
Project Manager: *Melissa M. Leick*
Senior Production Supervisor: *Sherry L. Kane*
Associate Design Coordinator: *Brenda A. Rolwes*
Cover Designer: *Studio Montage, St. Louis, Missouri*
(USE) Cover Images: *Factory Five Racing, Inc./Image-Tec*
Compositor: *Karen Fleckenstein, Fleck's Communications, Inc.*
Typeface: *10/12 Giovanni; Gill Sans Condensed*
Printer: *Quebecor World Dubuque, IA*

All credits appearing on page or at the end of the book are considered to be an extension of the copyright page.

Library of Congress Cataloging-in-Publication Data

Howard, William E. (William Edward), 1957–
  Introduction to solid modeling using solidworks 2009 / William E. Howard, Joseph C. Musto.—5th ed.
    p. cm.—(The McGraw-Hill graphics series)
  Includes index.
  ISBN 978-0-07-337540-3—ISBN 0-07-337540-3 (hard copy : alk. paper) 1. Computer graphics.
2. SolidWorks. 3. Engineering models. 4. Computer-aided design. I. Musto, Joseph C. II. Title.
  T385.H75 2010
  620'.00420285536—dc22
                        2009012750

www.mhhe.com

About the Authors

Ed Howard is an Assistant Professor in the Department of Engineering at East Carolina University. Prior to joining ECU, Ed taught for nine years at Milwaukee School of Engineering, where he taught classes in mechanics, finite element analysis, solid modeling, and composite materials. He was also Director of the Mechanical Engineering Technology program. He holds a BS in Civil Engineering and an MS in Engineering Mechanics from Virginia Tech, and a Ph.D. in Mechanical Engineering from Marquette University.

Ed worked in design, analysis, and project engineering for 14 years before beginning his academic career. He worked for Thiokol Corporation in Brigham City, UT, Spaulding Composites Company in Smyrna, TN, and Sta-Rite Industries in Delavan, WI. He is a registered Professional Engineer in Wisconsin.

Joe Musto is a Professor in the Mechanical Engineering Department at Milwaukee School of Engineering, where he teaches in the areas of machine design, solid modeling, and numerical methods. He holds a BS degree from Clarkson University, and both an M.Eng. and Ph.D. from Rensselaer Polytechnic Institute, all in mechanical engineering. He is a registered Professional Engineer in Wisconsin.

Prior to joining the faculty at Milwaukee School of Engineering, he held industrial positions with Brady Corporation (Milwaukee, WI) and Eastman Kodak Company (Rochester, NY). He has been using and teaching solid modeling using SolidWorks since 1998.

Joe and Ed, together with Rick Williams of East Carolina University, are the authors of *Engineering Computations: An Introduction using MatLab® and Excel®*, part of the McGraw-Hill "Best" Series.

CONTENTS

SPECIAL
FEATURES

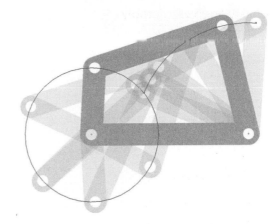

DESIGN INTENT

FUTURE STUDY

PREFACE

As design engineers and engineering professors, the authors have witnessed incredible changes in the way that products are designed and manufactured. One of the biggest changes over the past 20 years has been the development and widespread usage of solid modeling software. When we first saw solid modeling, it was used only by large companies. The cost of the software and the powerful computer workstations required to run it, along with the complexity of using the software, limited its use. As the cost of computing hardware dropped, solid modeling software was developed for personal computers. In 1995, the SolidWorks Corporation released SolidWorks®95, the first solid modeling program written for the Microsoft Windows operating system. Since then, the use of solid modeling has become an indispensable tool for almost any company, large or small, that designs a product.

The '33 Hot Rod featured on the cover is the product of Factory Five Racing, the world's leading manufacturer of component kit cars. Founded in 1995 by David and Mark Smith, Factory Five has made SolidWorks a key component of their revolution of the kit car industry. Factory Five engineers use SolidWorks exclusively for the design and manufacturing of the chassis, suspension, and components for their full line of cars.

SolidWorks software drives product development from concept to material management and final production tooling. The use of SolidWorks allows the engineers to design a kit that can perform better than the original car, while keeping the essence intact. Factory Five has received numerous awards, including Best in Show at SEMA, the country's largest automotive show held annually in Las Vegas.

Motivation for this Text

When we saw a demonstration of the SolidWorks software in 1998, we were both instantly hooked. Not only was the utility of the soft-

---

SolidWorks is a registered trademark of SolidWorks Corporation.

ware obvious, but the program was easy to learn and fun to use. Since then, we have shared our enthusiasm with the program with hundreds of students in classes at Milwaukee School of Engineering and East Carolina University, in summer programs with high school students, and in informal training sessions. Most of the material in this book began as tutorials that we developed for these purposes. We continue to be amazed at how quickly students at all levels can learn the basics of the program, and by the sophisticated projects that many students develop after only a short time using the software.

While anyone desiring to learn the SolidWorks program can use this book, we have added specific elements for beginning engineering students. With these elements, we have attempted to introduce students to the design process and to relate solid modeling to subjects that most engineering students will study later. We hope that the combination of the tutorial style approach to teaching the functionality of the software together with the integration of the material into the overall study of engineering will motivate student interest not only in the SolidWorks software but in the profession of engineering.

Philosophy of This Text

The development of powerful and integrated solid modeling software has continued the evolution of computer-aided design packages from drafting/graphical communication tools to full-fledged engineering design and analysis tools. A solid model is more than simply a drawing of an engineering component; it is a true virtual representation of the part, which can be manipulated, combined with other parts into complex assemblies, used directly for analysis, and used to drive the manufacturing equipment that will be used to produce the part.

This text was developed to exploit this emerging role of solid modeling as an integral part of the engineering design process; while proficiency in the software will be achieved through the exercises provided in the text, the traditional "training" exercises will be augmented with information on the integration of solid modeling into the engineering design process. These topics include:

- The exploitation of the parametric features of a solid model, to not only provide an accurate graphical representation of a part but also to effectively capture an engineer's design intent,

- The use of solid models as an analysis tool, useful for determining properties of components as well as for virtual prototyping of mechanisms and systems,

- The integration of solid modeling with component manufacturing, including the generation of molds, sheet metal patterns, and rapid prototyping files from component models.

Through the introduction of these topics, students will be shown not only the powerful modeling features of the SolidWorks program, but also the role of the software as a full-fledged integrated engineering design tool.

The Use of This Text

In this fifth edition of the text, we have completely updated the tutorials to reflect the new user interface and features of the SolidWorks 2009 software. The chapter-long tutorials introduce both basic concepts in solid modeling (such as part modeling, drawing creation, and assembly modeling) and more advanced applications of solid modeling in engineering analysis and design (such as mechanism modeling, mold creation, sheet metal bending, and rapid prototyping). Each tutorial is organized as "keystroke-level" instructions, designed to teach the use of the software. For easy reference, a guide to these tutorials is shown on the inside front cover.

While these tutorials offer a level of detail appropriate for new professional users, this text was developed to be used as part of an introductory engineering course, taught around the use of solid modeling as an integrated engineering design and analysis tool. Since the intended audience is undergraduate students new to the field of engineering, the text contains features that help to integrate the concepts learned in solid modeling into the overall study of engineering. These features include:

- *Design Intent Boxes:* These are intended to augment the "keystroke-level" tutorials to include the rationale behind the sequence of operations chosen to create a model.
- *Future Study Boxes:* These link the material contained in the chapters to topics that will be seen later in the academic and professional careers of new engineering students. They are intended to motivate interest in advanced study in engineering, and to place the material seen in the tutorials within the context of the profession.

While these features are intended to provide additional motivation and context for beginning engineering students, they are self-contained, and may be omitted by professionals who wish to use this text purely for the software tutorials.

The Organization of This Text

The organization of the chapters of the book reflects the authors' preferences in teaching the material, but allows for several different options. We have found that covering drawings early in the course is helpful in that we can have students turn in drawings rather than parts as homework assignments. The eDrawings feature, which is covered in Chapter 2, is especially useful in that eDrawings files are small (easy to e-mail), self-contained (not linked to the part file), and can be easily marked up with the editing tools contained in the eDrawings program.

The flowchart illustrates the relations between chapters, and can be used to map alternative plans for coverage of the material. For example, if it is desired to cover assemblies as soon as possible (as might be desired in a course that includes a project) then the chapters can be covered in the order 1-3-6-7-2-8, with the remaining chapters covered in any order desired. An instructor who prefers to cover parts, assemblies, and drawings in that order may cover the chapters in the order 1-3-4-5-6-7-2-8 (skipping section 4.4 until after Chapter 2 is covered), again with the remaining chapters covered in any order.

Chapters 9 and 10 may be omitted in a standard solid modeling course; however, they can be valuable in an introductory engineering course. Engineering students will almost certainly find use at some point for the 2-D layout and vector mechanics applications introduced in these chapters. Chapter 13 is intended to wrap up a course with a discussion of how solid modeling is used as a tool in the product development cycle.

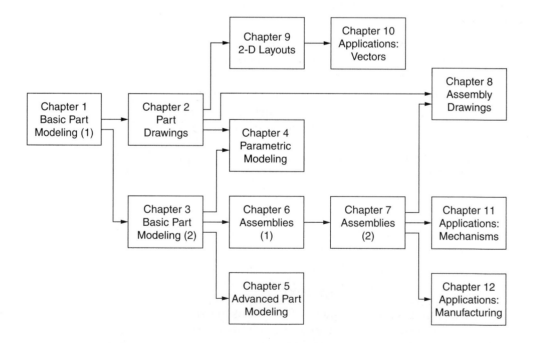

SolidWorks® Student Design Kit Included

The SolidWorks® Student Design Kit 2009–2010 can be downloaded from the text website at *www.mhhe.com/howard2009*. The student design kit includes a 150-day license of the following:

- *SolidWorks® 2009:* the standard in 3-D mechanical design software
- *SolidWorks® eDrawings® 2009:* a great tool for sharing design information
- *DWGEditor®:* a handy tool for viewing and editing 2-D drawings
- *SolidWorks® FloXpress™* and *SolidWorks® SimulationXpress:* powerful design validation tools

Online Resources for Instructors

Additional recourses are available on the web at *www.mhhe.com/howard2009*. Included on the website are tutorials for three popular SolidWorks Add-Ins, SolidWorks® Simulation, SolidWorks® Motion™ and Photo-Works™, and the book figures in PowerPoint format. Instructors can also access PowerPoint files for each chapter and model files for all tutorials and end-of-chapter problems (password-protected; contact your McGraw-Hill representative for access).

Acknowledgments

We are grateful to our friends at McGraw-Hill for their support and encouragement during this project. In particular, we have enjoyed working with Lora Neyens, our editor, and Bill Stenquist, the sponsoring editor. Karen Fleckenstein of Fleck's Communications, Inc. created the page design and did the page layouts. Brenda Rolwes created the cover design. Also, thanks to Tim Maruna, who encouraged us to initiate this project.

At SolidWorks Corporation, Marie Planchard and Christine Morse have provided continuous support for the project. The authors are also appreciative of the support of our SolidWorks resellers, Computer Aided Technology, Inc. and TriMech Solutions.

We also want to thank the reviewers whose comments have undoubtedly made the book better.

At the Milwaukee School of Engineering, many of our students and colleagues used early versions of the manuscript and materials that eventually became this text. We thank them for their patience and helpful feedback along the way.

Ed Howard
Joe Musto

PART ONE

Learning SolidWorks®

CHAPTER 1

Basic Part Modeling Techniques

Introduction

Solid modeling has become an essential tool for most companies that design mechanical structures and machines. Just 15 years ago, this would have been hard to imagine. While 3-D modeling software existed, it was very expensive and required high-end computer workstations to run. An investment of $50,000 or more was required for every workstation with software, not including training of the operator. As a result, only a few industries used solid modeling, and the trained operators tended to work exclusively with the software. The dramatic performance improvements and price drops of computer hardware, along with increased competition among software vendors, have significantly lowered the cost barrier for companies to enter the solid modeling age. The software has also become much easier to use, so that engineers who have many other job functions can use solid modeling when required without needing to become software specialists. The SolidWorks®[1] program was among the first solid modeling programs to be written exclusively for the Microsoft Windows environment. Since its initial release in 1995, it has been adopted by thousands of companies worldwide. This text is laid out as a series of tutorials that cover most of the basic features of the SolidWorks® program. Although these tutorials will be of use to anyone desiring to learn the software, they are written primarily for freshman engineering students. Accordingly, topics in engineering design are introduced along the way. "Future Study" boxes give a preview of coursework that engineering students will encounter later, and relate that coursework to the solid modeling tutorials. In this first chapter, we will learn how to make two simple parts in SolidWorks.

Chapter Objectives

In this chapter, you will:

- be introduced to the role of solid modeling in engineering design,

- learn how to create 2-D sketches and create 3-D extruded and revolved geometry from these sketches,

- use dimensions and relations to define the geometry of 2-D sketches,

- add fillets, chamfers, and circular patterns of features to part models,

- learn how to modify part models, and

- define the material and find the mass properties of part models.

---

[1] SolidWorks is a registered trademark of the SolidWorks Corporation, 300 Baker Avenue, Concord, MA 01742.

1.1   Engineering Design and Solid Modeling

The term *design* is used to describe many endeavors. A clothing designer creates new styles of apparel. An industrial designer creates the overall look and function of consumer products. Many design functions concentrate mainly on aesthetic considerations—how the product looks, and how it will be accepted in the marketplace. The term *engineering design* is applied to a process in which fundamentals of math and science are applied to the creation or modification of a product to meet a set of objectives.

Engineering design is only one part of the creation of a new product. Consider a company making consumer products, for example bicycles. A marketing department determines the likely customer acceptance of a new bike model and outlines the requirements for the new design. Industrial designers work on the preliminary design of the bike to produce a design that combines functionality and styling that customers will like. Manufacturing engineers must consider how the components of the product are made and assembled. A purchasing department will determine if some components will be more economical to buy than to make. Stress analysts will predict whether the bike will survive the forces and environment that it will experience in service. A model shop may need to build a physical prototype for marketing use or to test functionality.

During the years immediately following World War II, most American companies performed the tasks described above more or less sequentially. That is, the design engineer did not get involved in the process until the specifications were completed, the manufacturing engineers started once the design was finalized, and so on. From the 1970s through the 1990s, the concept of *concurrent engineering* became widespread. Concurrent engineering refers to the process in which engineering tasks are performed simultaneously rather than sequentially. The primary benefits of concurrent engineering are shorter product develop times and lower development costs. The challenges of implementing concurrent engineering are mostly in communications—engineering groups must be continuously informed of the actions of the other groups.

Solid modeling is an important tool in concurrent engineering in that the various engineering groups work from a common database: the solid model. In a 2-D CAD (Computer-Aided Design) environment, the design engineer produced sketches of the component, and a draftsman produced 2-D design drawings. These drawings were forwarded to the other engineering organizations, where much of the information was then duplicated. For example, a toolmaker created a tool design from scratch, using the drawings as the basis. A stress analyst created a finite element model, again starting from scratch. A model builder created a physical prototype by hand from the drawing parameters. With a solid model, the tool, finite element model, and rapid prototype model are all created directly from the solid model file. In addition to the time savings of avoiding the steps of recreating the design for the various functions, many errors are avoided by having everyone working from a common database. Although 2-D drawings are usually still required, since they are the best way to document dimensions and tolerances, they are linked directly to the solid model and are easy to update as the solid model is changed.

A mechanical engineering system (assembly) may be composed of thousands of components (parts). The detailed design of each component is important to the operation of the system. In this chapter, we will step through the creation of simple components. In future chapters, we will learn how to make 2-D drawings from a part file, and how to put components together in an assembly file.

1.2 Part Modeling Tutorial: Flange

This tutorial will lead you through the creation of a simple solid part. The part, a flange, is shown in **Figure 1.1** and is described by the 2-D drawing in **Figure 1.2**.

FIGURE 1.1

FIGURE 1.2

Begin by clicking the SolidWorks icon. Click on the New icon, as shown in **Figure 1.3**. If the Units and Dimension Standard box appears, as shown in **Figure 1.4**, select "IPS" as the units and "ANSI" as the standard. Click OK.

FIGURE 1.3

FIGURE 1.4

The Units and Dimension Standard box only appears the first time SolidWorks is opened. The selections become the default values for all new files. In this chapter, we will see how to set these values for individual files.

A dialog box will appear, from which you are to specify whether we are creating a part, an assembly, or a drawing (see **Figure 1.5**).

FIGURE 1.5

Click OK to accept Part as the type of file.

Before we begin modeling the flange, we will establish a consistent setup of the SolidWorks environment. The default screen layout is shown in **Figure 1.6**. The graphics area occupies most of the screen. The part, drawing, or assembly will be displayed in this area. There is a main menu bar at the top of the screen. The menu bar contains a toolbar with several commonly-used tools, and the name of the part is displayed on the menu bar.

FIGURE 1.6

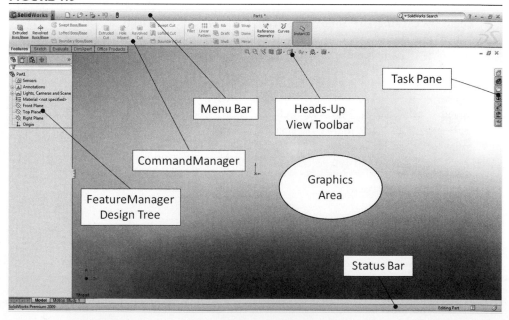

Note that if you pass the cursor over the SolidWorks button in the menu bar, a "fly-out" menu is displayed, as shown in **Figure 1.7**. The fly-out feature is designed to save room on the screen. However, since we will be using the menu often, we will disable the fly-out so that the menu is always displayed.

FIGURE 1.7

Move the cursor over the SolidWorks button to display the menu. Click on the pushpin icon at the right side of the menu, as shown in Figure 1.8, to lock the display of the menu.

FIGURE 1.8

The CommandManager contains most of the tools that you will use to create parts. When working in the part mode, there are two categories of tools that we will use extensively: Sketch tools used in creating 2-D sketches, and Features tools used to create and modify 3-D features. Clicking on the Sketch and Features tabs at the bottom of the CommandManager, as shown in **Figure 1.9**, changes the tools on the CommandManager to those of the selected group. By default, there are several other groups available besides the Sketch and Features groups. To simplify the interface, we will hide these groups for now.

FIGURE 1.9

Right-click on one of the CommandManager tabs. A list of available groups is displayed, with a check mark shown beside each active group (Figure 1.10). Click on any of the active groups other than Features and Sketch. This will clear the check mark and turn off the display of that group. Repeat until only the Features and Sketch groups remain active.

FIGURE 1.10

At the right side of the screen is the Task Pane. The Task Pane is a fly-out interface for accessing files and online resources. In keeping with our goal of a simple interface, we will turn off the Task Pane.

FIGURE 1.11

From the menu, select View: Task Pane, as shown in Figure 1.11.

This will toggle off the display of the Task Pane. If you want to turn it on later, simply select View: Task Pane again from the menu.

At the bottom of the screen is the Status Bar. When you move the cursor over any toolbar icon or menu command, a message on the left side of the Status Bar describes the command. Other information appears at the right side of the Status Bar, such as whether or not a sketch is completely defined. Although the display of the Status Bar can be toggled off and on from the View menu, we recommend leaving it on.

FIGURE 1.12

Just to the left of the drawing area is the FeatureManager Design Tree. The steps that you will execute to create the part will be listed in the FeatureManager. This information is important when the part is to be modified. When you open a new part, the FeatureManager lists an origin and three predefined planes (Front, Top, and Right), as shown in **Figure 1.12**. As you select each plane with your mouse, the plane is highlighted in the drawing area. We can create other planes as needed, and will do so later in this tutorial.

At the top of the graphics area is the Heads-Up View Toolbar. This toolbar contains many options for displaying your model. We will explore these options later in this tutorial.

We will now set some of the program options.

FIGURE 1.13

Select the Options tool from the menu bar toolbar, as shown in Figure 1.13. (You can also access the options from the menu, by selecting Tools: Options.)

The dialog box contains settings for both the system and for the specific document that is open.

FIGURE 1.14

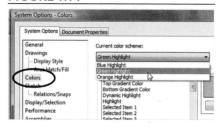

Under the System Options tab, choose Colors and change the color scheme to "Green Highlight," as shown in Figure 1.14.

Since this change was made to the System options, it will remain in effect for future SolidWorks sessions. The changes below, which will be made to the Document Properties, will apply only to the current part model.

Under the Document Properties tab, select Drafting Standard. Select ANSI from the pull-down menu, as shown in Figure 1.15.

FIGURE 1.15

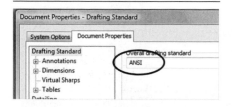

ANSI is the American National Standards Institute, an organization that formulates and publishes the standard drawing practices used by most companies in the United States. European companies are more likely to use the standards of ISO, the International Organization for Standardization.

Under the Document Properties tab, select Grid/Snap and make sure that the "Display grid" box is checked, as shown in Figure 1.16. Also under the Document Properties tab, select Units. Select IPS (Inch, pound, second) as the unit system. Use the pull-down menu under Decimals to set the number of decimal places for length units to 3 (.123), as shown in Figure 1.17. For Angle, set the number of decimal places to None, as shown in Figure 1.18.

FIGURE 1.16

FIGURE 1.17

FIGURE 1.18

Type	Unit	Decimals	F
Basic Units			
Length	inches	.123	
Dual Dimension Length	millimeters	.12	
Angle	degrees	None	

Note that there are "Dual Dimensions" units that can be set in the Units options. For some drawings, you may want to show dimensions in both US units (inches) and SI units (millimeters). Since we will not use dual dimensions for this part, it is not necessary to change the default settings.

Click OK to close the dialog box.

Any of the options just set can be changed at any time during the modeling process. Later in this chapter, we will learn how to create a template that allows us to begin a new part with our preferred settings in place.

FIGURE 1.19

We start the construction of the flange by sketching a circle and extruding it into a 3-D disk.

Select the Front Plane by clicking on it in the FeatureManager Design Tree, as shown in Figure 1.19.

The Front Plane will be highlighted in green. The color green indicates that an item is the currently selected item (since we chose the "Green Highlight" color scheme).

Begin a sketch by selecting the Sketch tab of the CommandManager, and then the Sketch Tool, as shown in Figure 1.20.

FIGURE 1.20

Note that when you selected the Front plane, a pop-up menu appeared that allowed you to open a sketch on that plane, as shown in **Figure 1.21**. The SolidWorks program has many of these context-sensitive menus built in. As you become proficient with the program, you may find many of these built-in shortcuts to be handy. Because this book is directed toward new users, we will mostly use commands from the CommandManager.

FIGURE 1.21

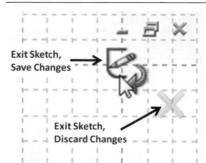

When you open a sketch, a grid pattern appears, signifying that you are in the sketching mode. Also, Exit Sketch icons appear in the upper-right corner of the screen, as shown in **Figure 1.22**.

FIGURE 1.22

Exit Sketch, Save Changes

Exit Sketch, Discard Changes

Select the Circle Tool from the Sketch group of the CommandManager, as shown in Figure 1.23.

FIGURE 1.23

When selecting any tool which has a pull-down menu (designated by the down arrow to the right of the icon), use caution to be sure that you are selecting the proper tool. In the case of the Circle Tool, there are two possible methods for defining the circle: by the center point and a point on the perimeter, or by three points on the perimeter. Clicking on the down arrow displays these options, so that the proper tool can be selected (**Figure 1.24**). By default, the option for defining the circle by locating the center point and a point on the perimeter is selected by clicking on the Circle Tool without accessing the pull-down menu. However, if the last option selected was to define the circle by three points on the perimeter, then that option becomes the default for the next selection. When that occurs, the icon shown for the Circle Tool will change, as shown in **Figure 1.25**. Because many of the icons are similar and are very small, you should use caution with tools that have pull-down menus.

FIGURE 1.24

FIGURE 1.25

FIGURE 1.26

You can check to see that you have selected the proper tool by looking at the PropertyManager, which appears in the area where the FeatureManager is normally shown whenever a tool is activated or an object is selected. The PropertyManager now shows the two alternative methods for defining a circle (**Figure 1.26**). If we selected the wrong tool accidentally, then we can change the method for defining the circle in the PropertyManager.

In the PropertyManager, make sure that the icon representing the first construction method is selected, as shown in Figure 1.26. If it is not, then click it to select it.

Notice as you move the cursor into the drawing area that it changes appearance into a pencil with a circle next to it, as shown in **Figure 1.27**. This lets you know that the Circle Tool is active.

FIGURE 1.27

Move the cursor toward the origin until a red dot appears at the origin, as shown in Figure 1.28; this indicates that you will snap to an existing point (in this case the origin) when you click with the mouse. Also, note the small icon next to the origin that signifies a coincident relation: the origin and the center point of the circle will share the same location.

A snap adds a relation to the positions of two entities. In this example, when you "snap" to the origin, the circle will be centered at the exact coordinates of $x = 0$ and $y = 0$. The relation added when one entity is created by snapping to another can be edited later, if desired. The addition of a snap automatically is a nice feature of the SolidWorks program: snaps are intuitive. It is not necessary to enter the numerical coordinates of the center of the circle.

FIGURE 1.28

With the center point highlighted as in Figure 1.28, click the left mouse button to place the center of the circle at the origin. Drag the mouse outward to create a circle, as shown in Figure 1.29. Click the left mouse button again to define a point on the perimeter and create the circle.

The circle will appear in green, indicating that it is the currently selected item.

FIGURE 1.29

Click the Esc key to close the Circle Tool and deselect the circle just drawn.

The circle should now appear in blue. In the Status Bar at the bottom of the screen, notice that "Under Defined" appears. This is because we have not set the diameter of the circle yet. When a sketch does not contain enough dimensions and/or relations to define its size and position in space, it is said to be under defined, and is denoted by blue entities.

Other possible conditions of the sketch are "Fully Defined," when the sketch contains exactly enough dimensions and/or relations to define its size and position in space (denoted by black entities), and "Over Defined," where the sketch has at least one dimension or relation that contradicts or is redundant to the other dimensions and relations (denoted by red entities). Over defined sketches should be avoided.

FIGURE 1.30

Select the Smart Dimension Tool from the CommandManager, as shown in Figure 1.30. Now click with the left mouse button anywhere on the circle. A

FIGURE 1.31

FIGURE 1.32

FIGURE 1.34

FIGURE 1.35

FIGURE 1.36

dimension will be added to the diameter of the circle. Drag the dimension to a convenient location, as shown in Figure 1.31. When the dimension is where you want to put it, left-click again to place the dimension.

A dialog box prompting for the value of this dimension will be displayed, as shown in Figure 1.32. Enter "5.5" in the box. (You don't need to enter "in" since that is the default unit, set earlier). Press the Enter key or click on the check mark to update the dimension.

Notice that the circle is redrawn to the correct dimension, as shown in **Figure 1.33**. The dimension in inches is displayed, and the circle is black. Notice at the bottom of the screen in the Status Bar that the sketch is now Fully Defined. (Note: If we had not snapped to the origin for the circle's center, the sketch would still be under defined because the circle would not be located in space.)

Use the Esc key to turn off the Smart Dimension Tool. Now if you double-click on the dimension value, the dialog box reappears and you can change the dimension. Try this, and then use the Undo Tool from the main menu toolbar, shown in Figure 1.34, to return the dimension to 5.5 inches.

FIGURE 1.33

Next to the center of the circle, an icon shows that a constraint is applied. By moving the cursor over the constraint icon, details about the constraint can be viewed, as shown in **Figure 1.35**. Constraints can be deleted by clicking on the constraint icon to select it, and then pressing the Delete key. The display of sketch relations can be toggled on and off by selecting View from the main menu and clicking on Sketch Relations. Some experienced users prefer to not show the relations because they can cause a sketch to appear cluttered, but new users are advised to keep the relation display turned on.

Now we are ready to turn this 2-D sketch into a 3-D part with the Extruded Boss/Base Tool.

Select the Features tab of the CommandManager, and the Extruded Boss/Base Tool, as shown in Figure 1.36.

The base feature is the first solid feature created. Any subsequent solid features are called bosses. Note that the view of the part changes to display a 3-D preview of the extruded solid. On the left side of the screen, the PropertyManager is now active. The PropertyManager allows the properties of the selected entity to be viewed and edited. There are several options available for the extrusion, including adding draft (taper) to the part, but for now we only need to adjust the depth of the extrusion.

Set the depth of the extrusion to 0.75 inches, as shown in Figure 1.37. Click Enter, and the preview in the graphics area will be updated to reflect the new thickness, as shown in Figure 1.38. Click on the check mark (OK) in the PropertyManager, as shown in Figure 1.39, and the circle is extruded into a solid disk, as shown in Figure 1.40.

FIGURE 1.37

FIGURE 1.38

FIGURE 1.39

FIGURE 1.40

Now that we have a solid part, we can examine the functions of the viewing tools. The viewing tools are located on the Heads-Up View Toolbar at the top of the graphics area. The default configuration of the Heads-Up Toolbar is shown in **Figure 1.41**. The Zoom to Fit Tool adjusts the zoom so that the entire model can be viewed. The Zoom to Area Tool allows a viewing window to be selected by dragging out an area on the screen. The Previous View Tool returns the view orientation and zoom level to the configuration prior to the most recent change of view.

FIGURE 1.41

FIGURE 1.42

The View Orientation Tool opens a menu of standard view options, as shown in **Figure 1.42**. The six principal views—Front, Back, Top, Bottom, Left, and Right—can be displayed by clicking on the appropriate icon. At the top of the menu are three pictorial views: Isometric, Trimetric, and Dimetric. In an isometric view, the view orientation is such that the angles between the displayed edges of a cube are equal, as shown in **Figure 1.43**. In a dimetric view, two of the angles are equal, and in a trimetric view all angles are different.

FIGURE 1.43

ISOMETRIC   DIMETRIC   TRIMETRIC

The Trimetric View in SolidWorks emphasizes the display of the front of the part, and is the default pictorial view. (As we will discuss later, the front view should be the view that is most descriptive of the part.) At the bottom of the menu are tools for displaying multiple views in separate windows on the screen. We will demonstrate their use later in this chapter. The Normal To Tool aligns the view to be perpendicular to a selected plane or surface. This tool is useful when sketching in a plane that is not perpendicular to any of the principal views.

The Display Style Tool opens a pull-down menu of options for displaying the model, as shown in **Figure 1.44**. There are two shaded modes, with or without the edges shown by lines, and three wireframe modes, with hidden edges removed, shown as dashed lines, or shown as solid lines. Usually, we work with one of the shaded modes. For some operations, displaying the model in wireframe mode is preferred.

FIGURE 1.44

The Hide/Show Items Tool allows you to toggle on or off the display of several items, such as the origin, planes, axes, etc. For now, we will skip over this tool, and will explore its use later in this chapter.

FIGURE 1.45

The Apply Scene Tool allows you to select backgrounds and lighting options from a pre-defined menu, as shown in **Figure 1.45**. In this book, we will use the Plain White scene. Some of the scenes contain background graphics rather than just colors. However, the display of these graphics requires a high-end graphics card. On most computers, only the colors and lighting will be changed by selecting a new scene. A SolidWorks Add-In program, PhotoWorks, can also be used to produce photo-realistic display images of the model in various scenes, even with computers without high-end graphics cards.

FIGURE 1.46

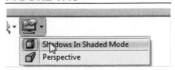

The View Settings Tool, shown in **Figure 1.46**, allows you to add shadows to either of the shaded modes. Also, the model can be shown in perspective mode. In perspective mode, sight lines converge at a single point (such as your eye), producing a more realistic view. However, most engineering views are produced with parallel sight lines. Normally, we will leave the Perspective View option turned off.

Several other viewing tools can be accessed by right-clicking in the white space of the graphics area. The menu shown in Figure 1.47 is displayed. A particularly useful tool is the Rotate View Tool. After selecting this tool, you can hold down the left mouse button and move the mouse to rotate the model view so that you can see all sides of the model. The Pan Tool can be used to move the model around in the graphics area, again by clicking and holding the left button while dragging the mouse. Note that when using either of these tools (or the Dynamic Zoom, also available from this menu), that the tool remains active until it is turned off by pressing the Esc key.

FIGURE 1.47

While we will discuss customization of the SolidWorks environment more fully in Chapter 3, we will make a modification to the Heads-Up View Toolbar here. Because the Rotate View and Pan Tools are used often, we will make them available on the toolbar. Also, since the Trimetric View is the default pictorial view, we will add it to the toolbar so that we do not have to go through the View Orientation Tool pull-down menu to select it.

Right-click on any of the tools in the Heads-Up View Toolbar. A list of available tools is displayed, with the active tools checked, as shown in Figure 1.48. Check the Rotate View, Pan, and Trimetric View tools to select them. Click anywhere away from the menu to close it.

FIGURE 1.48

The Head-Up Toolbar now contains the selected tools, as shown in **Figure 1.49**.

FIGURE 1.49

Experiment with the zoom and viewing options. When finished, select a shaded solid display (either with or without edges displayed) and the Trimetric View.

Now we are ready to add to our part. The next feature we will add is the 2.75-inch diameter boss. We could sketch the circle to be extruded on the front or back face of the existing part, but instead we will create a new plane that is 2.25 inches away from the Front Plane. There are several reasons why we might want to define the part in this manner. One is that we may want to add draft, a slope to the sides of a feature that allows it to be extracted from a mold. If so, then we want our 2.75-inch dimension to apply at the top of the boss, allowing the diameter to get larger closer to its base.

FIGURE 1.50

Select the Reference Geometry Tool from the Features group of the CommandManager. From the menu that appears, select Plane (see Figure 1.50).

Note that the FeatureManager has been replaced in its usual position by the PropertyManager, where the parameters of the new plane will be defined. However, the FeatureManager is still visible as a "fly out" list to the right of the PropertyManager. By default, the FeatureManager is shown collapsed; that is, only the name of the part is shown. The full FeatureManager can be shown by clicking on the plus sign next to the part name.

Click the plus sign next to the part name (Part1) to expand the FeatureManager, as shown in Figure 1.51. Click on the Front Plane to select it. In the box defining the offset distance, enter 2.25, as shown in Figure 1.52. Click the check mark and the new plane, labeled Plane1, is created, as shown in Figure 1.53.

FIGURE 1.51

FIGURE 1.52

FIGURE 1.53

There are several options for creating a new plane. When we selected the Front Plane, the option for defining the new plane parallel to the selected plane was selected by default. Of course, the Reverse direction box could have been checked if we wanted the new plane to be behind the Front Plane.

Now with Plane1 selected (highlighted in green), click the Sketch tab of the CommandManager and select the Circle Tool, as shown in Figure 1.54.

FIGURE 1.54

Remember that there are two tools defining different methods for defining a circle, with the last tool used the one that is selected by clicking the icon. If you hold the cursor over the icon momentarily, the current tool and its description are displayed in the Tooltips popup. If the correct tool is current, then you can select it by clicking on the icon, without using the pull-down menu.

Also note that a sketch was opened when you selected the Circle Tool. When a plane or face is selected, choosing a tool from the Sketch group opens a new sketch.

From the View Orientation Tool, select the Normal To View, as shown in Figure 1.55.

FIGURE 1.55

Notice that in this case, the Normal To View is the same as the Front View.

Move the cursor to the origin and click to place the center of the circle at the origin. Drag out a circle, as shown in Figure 1.56, and click to complete the circle.

FIGURE 1.56

For clarity, most of the images in this text were produced with the grid display turned off. For new users, however, the grid display is a useful reminder that the sketching mode is active.

Select the Smart Dimension Tool from the Sketch group of the CommandManager, as shown in Figure 1.57. Add a 2.75-inch dimension to the diameter, as shown in Figure 1.58. Switch to the Trimetric View (Figure 1.59). Select the Extruded Boss/Base Tool from the Features group of the CommandManager (Figure 1.60).

FIGURE 1.57

FIGURE 1.58

FIGURE 1.59

FIGURE 1.60

DESIGN INTENT | Planning the Model

As we build the 2.75-inch diameter boss of our flange, we can choose from two existing planes/surfaces or construct a new plane. The choice of constructing a new plane in order to allow us to add draft to our part is an example of design intent. There are many definitions of design intent. Ours is:

Design intent is the consideration of the end use of a part, and possible changes to the part, when creating a solid model.

Throughout this book, we will identify examples of considering design intent when modeling.

FIGURE 1.61

As shown by the preview (**Figure 1.61**), the extrusion is going away from the base feature rather than toward it.

In the PropertyManager, click the Reverse Direction button so that the extrusion is directed toward the base (Figure 1.62). In the pull-down menu, select Up To Next as the type of extrusion, as shown in Figure 1.63. Click the check mark to complete the extrusion.

FIGURE 1.62

FIGURE 1.63

We can turn off the display of Plane1.

FIGURE 1.64

From the Hide/Show Items Tool, click on the View Planes icon to toggle off the display of planes, as shown in Figure 1.64. Also from the Hide/Show Items Tool, click on the View Origins icon.

Plane1 and the origin still exist in the model, but turning off the display of planes and origins results in a less cluttered model view, as shown in Figure 1.65.

It is a good idea to save your work periodically.

FIGURE 1.65

Choose File: Save from the main menu. Save the part with the name "Flange." The file type will be "sldprt."

Note that the new file name appears in the Menu Bar and at the top of the FeatureManager design tree.

We can change the color of the part at any time.

Right-click on the part name (Flange) in the FeatureManager, and choose Appearance: Appearance, as shown in Figure 1.66. Choose a new color for the part (see Figure 1.67). Light colors are recommended, as dark colors can make some features and selections difficult to see. Click the check mark in the PropertyManager to apply the new color.

FIGURE 1.66

FIGURE 1.67

You may notice that the circular edges appear to be a series of line segments rather than a smooth curve. You can smooth the appearance of the part by selecting a higher image quality.

Choose the Options Tool, as shown in Figure 1.68. Under the Document Properties tab, choose Image Quality. Move the slider bar to the right, as shown in Figure 1.69. Click OK.

FIGURE 1.68

The trade-off for higher image quality is that the speed of redrawing the screen will be reduced slightly. However, for simple parts this speed reduction will not be noticeable.

Now let's add the center hole. This time we will select a face to define our sketch plane. As you move the cursor over the front surface, notice that a square icon appears. This indicates that a surface will be selected when you click with the left mouse button. Similarly, a line icon indicates that an edge will be selected.

FIGURE 1.69

Move the cursor over the front face, so that the square icon appears, as shown in Figure 1.70. **Click to select this face; it will be highlighted in green. A pop-up tool-bar, called a Context Toolbar in SolidWorks, will appear. Click the Normal To View, as shown in** Figure 1.71.

FIGURE 1.70

FIGURE 1.71

(Note: if the Context Toolbar does not appear when you select the face, then select the Normal To or Front View from the Heads-Up View Toolbar instead. We will see how to control the appearance of the Context Toolbar in the next section of this chapter.)

FIGURE 1.72

Click the Sketch tab on the CommandManager and select the Circle Tool. Drag out a circle centered at the origin. Select the Smart Dimension Tool and dimension the circle diameter as 1.5 inches, as shown in Figure 1.72. **Now click the Features tab of the CommandManager and select the Extruded Cut Tool, as shown in** Figure 1.73. **Select the type as Through All in the PropertyManager, as shown in** Figure 1.74. **Click the check mark to complete the cut. The result of this operation is shown in** Figure 1.75.

We will now add the four bolt holes.

FIGURE 1.73

FIGURE 1.74

FIGURE 1.75

FIGURE 1.76

Select the surface shown in Figure 1.76. Switch to the Normal To View. Click the Sketch tab of the CommandManager, and select the Circle Tool. Drag out a circle centered at the origin. In the PropertyManager, check the "For construction" box, as shown in Figure 1.77.

Construction entities help you locate and size sketch parameters, and are indicated by dashed-dotted lines. The circle just drawn represents the bolt circle.

FIGURE 1.77

Select the Smart Dimension Tool. Add a 4.25-inch diameter dimension to the circle, as shown in Figure 1.78. Select the Circle Tool. Move the cursor to the top quadrant point on the construction circle, as shown in Figure 1.79. Note the red diamond that appears, along with the coincident and vertical relation icons. Drag out a circle, as shown in Figure 1.80. Select the Smart Dimension Tool and add a diameter dimension of 0.50 inches, as shown in Figure 1.81.

FIGURE 1.78

FIGURE 1.79

FIGURE 1.80

FIGURE 1.81

The sketch is fully defined, since the center of the circle just drawn has been located at the top quadrant point of the bolt circle.

Select the Extruded Cut Tool from the Features group of the CommandManager, and extrude a hole with a type of Through All. Click the check mark.

The first bolt hole is now in place, as shown in the trimetric view in **Figure 1.82**. Note that when we selected the Extruded Cut tool, only the small circle was used as the geometry of the cut. The bolt circle, because it was identified as construction geometry, was not included as a sketch entity to be extruded.

FIGURE 1.82

Notice that in the FeatureManager, all of our procedures are being recorded. The names of the features are not particularly descriptive; the four features that we have created so far were all created by extrusions, and so are named "Extrude1," "Extrude2," etc. To more easily identify features for later modifications, we can rename features.

Click once on Extrude4 in the FeatureManager to select and highlight the name. Click again to allow editing of the name. (Use two separate mouse clicks, not a double-click.) Type "Bolt Hole" to rename the feature, as shown in Figure 1.83. Press the Enter key to accept the new name.

We could create the other three holes separately, but it is easier to copy the single hole into a circular pattern. Also, since our design intent is for the holes to exist in a circular pattern, it makes sense to construct them that way. If we later change the diameter of the holes, the diameter of the bolt circle, or the number of holes, it will be easy to do if we have created them in a pattern.

FIGURE 1.83

Make sure that the first bolt hole is selected. Click the Features tab of the CommandManager, and click the arrow under the Linear Pattern Tool to reveal a menu of pattern tools, as shown in Figure 1.84. Choose the Circular Pattern Tool.

Select the cylindrical face shown in Figure 1.85. This causes the axis of the face to be set as the axis of rotation for the hole pattern. (Note that the outer face of the large

FIGURE 1.84

FIGURE 1.85

part of the flange or the inside face of the center hole could have been
selected instead, as all three of these faces have the same center axis.) In
the PropertyManager, check the "Equal spacing" box, which will cause the
angle to be changed to 360 degrees. Change the number of holes to 4 by
clicking the up arrow, as shown in Figure 1.86. Click the check mark to complete the pattern.

FIGURE 1.86

Now let's finish the flange by adding fillets to three of the sharp edges. A fillet is a feature that rounds off a sharp edge. Actually, a fillet is a rounded edge
created by adding material, while a round is created by removing material.
Fillets and rounds are created with the SolidWorks software by the same
command.

From the Features group of the CommandManager, select the Fillet Tool, as shown
in Figure 1.87. Select the three edges indicated in Figure 1.88 to be filleted. (Be
sure to see the line next to the cursor, as shown in Figure 1.89, to indicate that
an edge and not a face is being selected.) In the PropertyManager, enter the
radius as 0.25 inches, as shown in Figure 1.90. Check the Full preview box to see
the fillets that will be created, as shown in Figure 1.91. Click the check mark to
add the fillets, shown in Figure 1.92.

FIGURE 1.87

FIGURE 1.88

FIGURE 1.89

FIGURE 1.90

FIGURE 1.91

FIGURE 1.92

DESIGN INTENT | Selecting a Modeling Technique

The three fillets are added in this tutorial in a single step by selecting the three edges to be filleted within a single fillet command. With this method, only the first fillet is dimensioned. Another way to add the fillets is to close the Fillet Tool after each fillet is created, so that the fillets are created in three separate steps. The preferred method depends on how you wish to edit the fillet radii. If you want all of the fillets to always have the same radius, then the first method allows one value to be changed for all three fillets. If you prefer to edit the fillets separately, then the second method provides an editable dimension for each fillet.

Display of tangent edges is often undesirable. The display of tangent edges can be controlled from the Options menu.

Select the Options Tool. Under the System Options tab, under Display/Selection, choose Removed as the Part/Assembly tangent edge display option, as shown in Figure 1.93. Click OK.

The part should appear as in **Figure 1.94**.

FIGURE 1.93

FIGURE 1.94

Now we can add the chamfer to the center hole. A chamfer is a conical feature formed by removing material from an edge.

FIGURE 1.97

FIGURE 1.95

FIGURE 1.96

Select the arrow under the Fillet Tool in the CommandManager, and select the Chamfer Tool, as shown in Figure 1.95. Click on the edge shown in Figure 1.96 to select it as the edge to be chamfered. In the PropertyManager, set the chamfer parameters to 0.080 inches and 45 degrees, as shown in Figure 1.97, and click the check mark to finish.

FIGURE 1.98

The finished part is shown in **Figure 1.98**.

From the main menu, select File: Save. Leave the part file open for the next section, in which we will learn how to make modifications to the part.

1.3 Modifying the Flange

One of the main advantages of solid modeling is the ability to make changes easily. As we have observed, the FeatureManager has recorded all of the operations required to make the flange, as shown in **Figure 1.99**. If we click on the plus sign next to each feature, we see that the sketch associated with each feature is stored, as well.

FIGURE 1.99

Let's change the first item that we created by increasing the diameter of the base from 5.5 to 7 inches.

Right-click Sketch1 in the FeatureManager.

Notice that at the top of the menu, there is a Context toolbar including an Edit Sketch Tool, as shown in **Figure 1.100**. While these Context toolbars are designed to make selections faster, for new users they may be more difficult to use because the name of the tool is not displayed unless you hold the cursor over it momentarily. We will change a setting so that the commands displayed on the Context toolbar are shown within the menu, instead.

FIGURE 1.100

Click the arrow beside the Options Tool and select Customize, as shown in Figure 1.101. Under the Toolbars tab, clear the check box labeled "Show in shortcut menu," as shown in Figure 1.102. Leave the "Show on selection" box checked. Click OK.

FIGURE 1.101

FIGURE 1.102

FIGURE 1.103

FIGURE 1.106

Right-click again on Sketch1 in the FeatureManager. Select Edit Sketch from the menu, as shown in Figure 1.103. Double-click the 5.5-inch dimension, and change it to 7.0 inches, as shown in Figure 1.104. When you close the sketch by clicking on the icon in the upper-right corner of the screen (see Figure 1.105), the part will be updated to the new dimension, as shown in Figure 1.106.

FIGURE 1.104

FIGURE 1.105

An even easier way to edit the sketch dimensions or the extrude depth is illustrated next.

Double-click the icon next to Extrude1 in the FeatureManager. All of the dimensions used to create the feature are displayed, as shown in Figure 1.107. The sketch dimensions are shown in black, while the feature dimensions (in this case the extrude depth) are shown in blue. Double-click the diameter dimension and change it back to 5.5 inches. Click the Rebuild Tool, as shown in Figure 1.108.

FIGURE 1.107

FIGURE 1.108

FIGURE 1.109

To add draft to the boss, select Extrude2 from the FeatureManager, right-click and select Edit Feature, as shown in Figure 1.109. In the PropertyManager, turn the draft on (see Figure 1.110) and set the angle to 3 degrees. Check the "Draft outward" box so that the boss increases in size as it is extruded. Click the check mark to finish.

FIGURE 1.110

The draft will be easier to see from a top or side view. You can show the Front, Top, and Right Views along with the current (Trimetric) view with the Four-View option. Before displaying the Four View window, we will set an option that controls how the views are displayed.

FIGURE 1.111

Select the Options Tool. Under the System Options tab, select Display/Selection. From the pull-down menu near the bottom of the dialog box, make sure that "Third Angle" is set, as shown in Figure 1.111.

The third-angle option displays the Front, Top, and Right Views, with the Front View in the lower left position (standard for most US drawings), while the first-angle option displays the Front, Top, and Left Views, with the Front View in the upper left position (standard for most European drawings).

Select the Four-View window from the View Orientation Tool of Heads-Up View Toolbar, as shown in Figure 1.112.

The drafted feature can be seen clearly in the Top and Right Views, as shown in Figure 1.113.

FIGURE 1.112

FIGURE 1.113

To revert to a single view, click in the window displaying the Trimetric View, and select Single View from the View Orientation Tool of the Heads-Up View Toolbar, as shown in Figure 1.114.

Finally, right-click on CirPattern1 in the FeatureManager and select Edit Feature. Change the number of holes from four to six, as shown in Figure 1.115.

FIGURE 1.114

FIGURE 1.115

The modified part is shown in **Figure 1.116**.

FIGURE 1.116

These last two changes illustrate the importance of considering design intent when modeling. If the first boss had been extruded from the base feature, then adding draft would have required us to change the diameter of the boss, calculating the diameter that will result in a 2.75-inch diameter at the top of the boss when draft is included. By sketching in a plane at the top of the boss, the critical 2.75-inch dimension can be maintained easily. Also, by constructing the holes as a circular pattern instead of individually, the number of holes could be modified easily.

Click the Undo Tool (Figure 1.117) to reverse the previous command.

There is another method of editing a part, called Instant 3D, that we will not use in this book, but is worth mentioning.

FIGURE 1.117

Click the Features Tab of the CommandManager. If the Instant 3D Tool is not currently active (as indicated by a "depressed" icon), click to turn it on, as shown in Figure 1.118. Click on the front face of the model. Notice that an arrow perpendicular to the face appears. Click and drag on the arrow, as shown in Figure 1.119, to change the overall depth of the part.

FIGURE 1.118

FIGURE 1.119

While this method of editing the model is very useful in more "free form" design, we recommend that new users refrain from using the Instant 3D feature to avoid changing dimensions accidentally.

Click again on the Instant 3D Tool to turn it off. Close the part window by clicking on the X in the upper-right corner of the part window. Do not save the changes to the file.

1.4 Sketch Relations in Modeling

In order to fully define a sketch, we use both dimensions and relations. In the previous section, our first sketch consisted of a circle drawn in the Front Plane. We defined the size of the circle with a diameter dimension, but we used a relation to define its position. This coincident relation was created automatically when we snapped the center point to the origin. Later, we located the center of the first bolt hole by adding a coincident relation between the center of the circle defining the hole and a quadrant point on the construction circle defining the bolt circle. In this section, we will learn more about adding and deleting relations.

FIGURE 1.120

Automatic relations can be toggled on or off from the Tools: Sketch Settings menu, as shown in **Figure 1.120**. There will be very few instances in which new users will want to turn off automatic relations; in this text, we will assume that they are always turned on. To illustrate the use of automatic relations, consider the line shown in **Figure 1.121**. Suppose that we want to draw a new line beginning at the midpoint of the original line. When the Line Tool is active, and the tip of the "marker" icon is placed near the first line, that line is highlighted and the midpoint of the line is displayed, as shown in **Figure 1.122**. Also, note the coincident icon

FIGURE 1.121   **FIGURE 1.122**

that appears. If we were to click to begin the new line at this point, the new line would have an endpoint on the original line, but its exact location along the line would still be undefined. If we move the tip of the marker closer to the midpoint of the line before clicking, then only the midpoint will be highlighted and a mid-point icon will be displayed, as shown in **Figure 1.123**. Clicking here will begin the new line precisely at the midpoint of the original line. When we drag out a new line, guidelines that are parallel and perpendicular to the original line will appear on the screen. If the cursor is placed along the perpendicular line, as shown in **Figure 1.124**, then a perpendicular icon appears and clicking to end the line will

FIGURE 1.123   **FIGURE 1.124**

result in the addition of a perpendicular relation. The relations are shown on the screen by the icons next to the sketch entities that they apply to, as shown in **Figure 1.125**. The display of these icons can be toggled on and off from the Hide/Show Items menu of the Heads-Up View Toolbar, as shown in **Figure 1.126**. We recommend leaving the display of sketch relation icons turned on, although occasionally for very complex sketches, you may want to turn off their display for a less cluttered graphics area.

FIGURE 1.125

FIGURE 1.126

Moving the cursor over one of these icons displays information about that relation and highlights the affected sketch entities, as shown in **Figure 1.127**. Note that in the case of the midpoint relation, the original line is highlighted, but only the endpoint of the new line. A midpoint relation always exists between a line (or an arc) and a single point. If you hold your cursor over the perpendicular icon, then both lines are highlighted, as the perpendicular relation applies to the two lines.

FIGURE 1.127

Midpoint(Point3, Line1)

FIGURE 1.128

Relations can also be applied manually. For example, consider the lines shown in **Figure 1.128.** To add relations to existing sketch entities, we must first select those entities. The Ctrl key allows for multiple selections. To add the midpoint relations, we would click to select the endpoint of the line to the right, and, while holding the Ctrl key, click to select the other line. With these entities selected, a Context menu appears with possible relations, including a midpoint relation, as shown in **Figure 1.129**. Also, the PropertyManager displays any existing relations between the selected entities, and provides options for adding relations, as shown in **Figure 1.130**. After applying the relation, either clicking the check mark in the PropertyManager or clicking in an empty area of the graphics area de-selects the entities. Selecting both lines results in a Context menu that contains a perpendicular relation, as shown in **Figure 1.131**.

FIGURE 1.129

Make Midpoint

FIGURE 1.130

FIGURE 1.131

Make Perpendicular

Relations can also be applied to a single entity. In this case, some aspect of the entity's geometry is related not to another sketch entity, but rather to the pre-defined axes of the sketch plane. For example, if we select the left line, then we are presented options for making this line vertical or horizontal, or for fixing it in its current position in the sketch, as shown in **Figure 1.132**.

FIGURE 1.132

Often, it is necessary to delete a relation. For example, suppose that we want to change our sketch so that the two lines are oriented at 45 degrees relative to one another, rather than perpendicular. If we try to apply an angular dimension, then we get a message that doing so will overdefine the sketch, as shown in **Figure 1.133**. If we select the vertical line, then the PropertyManager displays all of the relations involving that line, as shown in **Figure 1.134**. A relation can be deleted by right-clicking and selecting delete, or by highlighting it and pressing the Delete key. With the perpendicular relation deleted, then the 45-degree angle can be added, as shown in **Figure 1.135**.

FIGURE 1.133   **FIGURE 1.134**   **FIGURE 1.135**

Relations are particularly useful for minimizing the number of dimensions required to fully define a sketch. For example, consider the T-section shown in **Figure 1.136**. Because this shape is to be symmetric, we would like to use only four unique dimensions to define its geometry. But to ensure symmetry, we need two more dimensions—one that forces the two "legs" to be equal in thickness, and the other that causes the upright portion to be centered. If we delete the two undesired dimensions, then portions of the geometry are allowed to "float," as shown in **Figure 1.137**. (It is often useful to click and drag entities of

FIGURE 1.136   **FIGURE 1.137**

an under defined sketch in order to determine what aspects of the geometry are not completely defined.) Selecting the two horizontal lines shown in **Figure 1.138** and applying an equal relation forces the upright portion of the geometry to be centered, eliminating the need for the previous 1.2-inch dimension. Another equal relation between a pair of vertical lines could be added to fully define the sketch, but an alternate approach is to add a collinear relation between the horizontal lines already selected, as shown in **Figure 1.139**. There are almost always multiple approaches to achieving the same desired geometry. The advantage of using relations instead of the two additional dimensions is seen when making changes to the sketch. As shown in **Figure 1.140**, any of the dimensions can be changed without changing our design intent that the section always be symmetric.

FIGURE 1.138

FIGURE 1.139

FIGURE 1.140

A list of the relation icons is shown in **Figure 1.141**. All are common in 2-D sketches except for the last two: the Pierce relation is used in multiple-sketch applications such as sweeps and lofts, and the Along X relation is used in 3-D sketches (there are similar Along Y and Along Z relations).

FIGURE 1.141

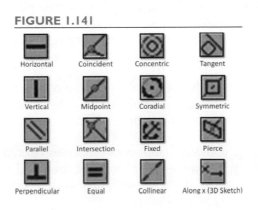

1.5 A Part Created with Revolved Geometry

The flange created earlier utilized extruded features. In this exercise, we will use *revolved* features to create the pulley shown in **Figure 1.142**. We will sketch features of the cross-section of the pulley, shown in **Figure 1.143**, and then revolve those features around a centerline to create solids and cuts. The final feature will be a keyway, detailed in **Figure 1.144**, which will be made with an extruded cut.

FIGURE 1.142

FIGURE 1.143

FIGURE 1.144

Open a new part.

You will notice that some of the changes made earlier to the SolidWorks interface, such as the number of tabs on the CommandManager and the tools on the Heads-Up Toolbar, are still in effect. Others, such as the background color, have reverted to the initial settings and need to be changed again. These document-specific settings are stored in the Parts *template*. We will edit the template so that we do not have to make these changes every time we begin a part.

Change the background color to Plain White from the Apply Scene Tool of the Heads-Up Toolbar. Select the Options Tool, and click the Document Properties tab. Change the drafting standard to ANSI. Under Dimensions, set the Primary precision to .123 (three decimal places). Select Grid/Snap, and check the box labeled "Display Grid." Select Units, and change the unit system to IPS, the number of decimals for length dimensions to .123, and the number of decimals for angles to None. Select Image Quality, and move the top slider bar toward the right (High).

Note that we have set the number of decimal places in two separate locations. The number of places can be changed in either location; by setting both to .123 we ensure that the template setting will be stored correctly.

Any of the other settings under the Document Properties tab can be stored in the template. For example, you may want to change the font used for dimensions or turn the display of the grid off (most of the figures in this book are made with a larger font and with the grid off for clarity).

From the main menu, select File: Save As. Change the file type to Part Templates (*.prtdot). Click on the file "Part.prtdot" to select it, as shown in Figure **1.145, and click Save.**

FIGURE 1.145

The next time you begin a new part, the settings that you just made will be effective. Note that when you changed the file type to Part Template, the working directory automatically changed to default directory for SolidWorks templates. This directory is located in the Program Files or Program Data directory on the drive containing your operating system. Therefore, it is not a good location to store your work. The next time that you save a document, the directory will default to the last one accessed; in this case the directory where the templates are saved. Make sure to change the directory to the one you want before saving any documents. We will note this when we save the pulley file later in this section.

FIGURE 1.146

Click on the Right Plane to select it, as shown in Figure 1.146. **From the Sketch group of the CommandManager, click the arrow beside the Line Tool and select the Centerline Tool, as shown in** Figure 1.147. **Draw a vertical centerline up from the origin, as shown in** Figure 1.148.

FIGURE 1.147
FIGURE 1.148

This centerline will allow us to take advantage of the symmetry of the cross-section.

FIGURE 1.149

Click the arrow beside the Rectangle Tool, and select the Center Rectangle Tool, as shown in Figure 1.149. **Click above the origin on the centerline, as shown in** Figure 1.150, **to set the center of the rectangle. Then drag out a corner of the rectangle, as shown in** Figure 1.151. **The size is not important, but keep the entire rectangle above the origin. Repeat to create a second rectangle above the first, as shown in** Figure 1.152.

FIGURE 1.150

FIGURE 1.151
FIGURE 1.152

x = 2.715, y = 0.491

x = 1.862, y = 0.517

Click the arrow beside the Rectangle Tool, and select the Corner Rectangle Tool, as shown in Figure 1.153. Move the cursor to the bottom line of the top rectangle (but not to a corner or midpoint), as shown in Figure 1.154, and click to place one corner of the rectangle on this line. Drag the rectangle down until the opposite corner is along the top line of the bottom rectangle, as shown in Figure 1.155. Click to complete the rectangle.

FIGURE 1.153

FIGURE 1.154

FIGURE 1.155

FIGURE 1.156

Select the Smart Dimension Tool, and click once on the top line of the top rectangle to create a linear dimension. Drag the dimension to the desired location and click to place it. Enter the value as 1.5 inches, as shown in Figure 1.156. Add the 2.0-inch dimension shown in Figure 1.157 to the bottom line. Press the Esc key to turn the Smart Dimension Tool off.

FIGURE 1.157

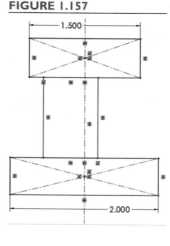

When we used the Center Rectangle Tool and placed the center points along the vertical centerline, those rectangles became symmetric about the centerline. However, to join the first two rectangles together, we used a Corner Rectangle, which is not centered on the centerline. Therefore, we need to add the symmetry of this rectangle manually.

Click on the two lines and the centerline shown in Figure 1.158 to select them, remembering to hold down the Ctrl key when making multiple selections. Click on Make Symmetric Icon from the pop-up menu, as shown in Figure 1.159 (or the Symmetric Icon in the PropertyManager).

FIGURE 1.158

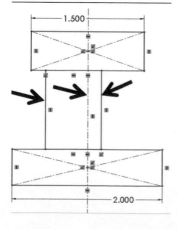

The middle rectangle should now be symmetric about the centerline, as shown in Figure 1.160.

FIGURE 1.159

FIGURE 1.160

Select the Smart Dimension Tool, and click on one of the vertical lines of the middle rectangle, and then on the other vertical line, creating a linear dimension. Drag the dimension to the desired location, and then click to place it. Enter the value as 0.25 inches, as shown in Figure 1.161.

Note that it is not necessary to hold down the Ctrl key when selecting multiple entities for the Smart Dimension command. If you click on a single entity and then click away from any other entity to place it, the Smart Dimension Tool creates a dimension from that entity (length of a line, diameter of a circle, radius of an arc). If you click on an entity and then click on a second entity, the Smart Dimension Tool will attempt to create a dimension relating the two entities. When we selected two parallel lines, the distance between the two lines was added as a dimension. Later, we will see that if we select two non-parallel lines, an angular dimension will be created.

If you refer back to **Figure 1.143**, you will see that the other dimensions used to define this cross section are diameter dimensions. In order to place the diameter dimensions, we need to establish the centerline, which will become the axis of revolution for the resulting solid.

Select the Centerline Tool, and drag a horizontal centerline from the origin, as shown in Figure 1.162. Select the Smart Dimension Tool. Click on the top line of the sketch and then on the horizontal centerline. Before clicking to place the dimension, note that if you place the dimension above the centerline, as in Figure 1.163, then the linear distance between the two (the radius) is defined. However, if you place the dimension *below* the centerline, as in Figure 1.164, then the linear dimension between the top line and a mirror image of that line below the centerline (the diameter) is defined. Click below the centerline, and enter the value of the dimension as 7.0 inches, as shown in Figure 1.165. Repeat to add the 5.0-, 1.75-, and 1.0-inch diameter dimensions shown in Figure 1.166.

The sketch should be fully defined.

FIGURE 1.161

FIGURE 1.162

FIGURE 1.163

FIGURE 1.164

FIGURE 1.165

FIGURE 1.166

DESIGN INTENT | Planning for Other Uses of the Model

When adding dimensions to the initial sketch of the pulley, we could specify some of the dimensions as either diameters or radii. For example, the 7-inch-diameter dimension that defines the overall size of the pulley could just as easily be entered as a 3.5-inch-radius dimension. However, if we plan to make a 2-D drawing of this part, then the diameter should be defined as a diameter on the drawing. By dimensioning the part in the same way that we will dimension the drawing, then dimensions can be imported directly from the part file. This prevents us from having to add dimensions manually or override a dimension's properties. When you are planning to make a 2-D drawing of a part; use the part in an assembly; or utilize the part file for another use such as conducting a stress analysis, creating a rapid prototyped model, or defining a tool path; consideration of the uses of the model will often influence the best way to create and define the part geometry.

From the Features group of the CommandManager, select the Revolved Boss/Base Tool, as shown in Figure 1.167.

Note that no preview is displayed on the screen as it was when we selected the Extrude Boss/Base tool earlier in the chapter. The reason for this is that there are two ambiguities in our sketch that must be defined:

1. There is more than one enclosed region within our sketch, so we need to define which of these regions will be revolved.
2. There are two centerlines in the sketch, so we need to define which centerline is the axis of revolution.

One way to approach the first ambiguity would be to use the Trim Entities Tool to remove the overlapping portions of the rectangles, so that the sketch consists of only a single closed contour. An easier way is to simply select the multiple contours. In the PropertyManager, note that the "Selected Contours" box is highlighted, as shown in Figure 1.168.

FIGURE 1.167

FIGURE 1.168

FIGURE 1.169

FIGURE 1.170

Move the cursor into the top rectangle, and click to select the rectangular region, as shown in Figure 1.169. Repeat for the other two rectangular regions, as shown in Figure 1.170. It may be necessary to zoom in to select the middle region.

Click in the Axis box in the PropertyManager to select it, as shown in Figure 1.171. Click on the horizontal centerline, as shown in Figure 1.172.

FIGURE 1.171

FIGURE 1.172

A preview will now be displayed, as shown in **Figure 1.173**. In the PropertyManager, we can change the number of degrees of the revolution if we want less than a fully revolved part. Since we want to revolve the section a full 360 degrees, we can accept the default value.

FIGURE 1.173

Click the check mark to complete the revolution.

The resulting solid part is shown in **Figure 1.174**.

FIGURE 1.174

Save the file with the name "Pulley." Make sure to change the file directory to the location where you want to save the file, since the default path will be to the directory containing the template files.

Select the Right Plane, and select Normal To from the pop-up toolbar, as shown in Figure 1.175. Select the Centerline Tool, and create vertical and horizontal centerlines from the origin, as shown in Figure 1.176.

FIGURE 1.175

FIGURE 1.176

DESIGN INTENT | Choosing the Initial Sketch Plane

The choice of the initial sketch plane for a part should be considered before beginning the first sketch. Standard practice is to orient the part so that the Front View provides the best visualization of the part of all of the principal views (Front, Back, Top, Bottom, Right, and Left). Therefore, the Trimetric View, which emphasizes the front of the part, provides the best pictorial (3-D) view. For the flange part created earlier in this chapter, sketching the initial circle in the Front Plane resulted in the proper orientation of the flange. As we will see in Chapter 2, orienting the part properly will make the creation of a multi-view drawing easier, as well.

In the case of the pulley, which is created by revolved features, the circular profile of the part is the most descriptive view, so we chose the Right Plane as the sketch plane for our initial sketch. We could have also sketched in the Top Plane and achieved the same result.

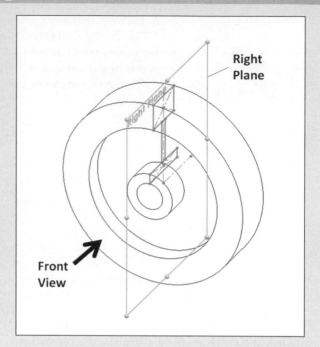

Once a part has been created in a certain orientation, it is difficult to re-orient it. There is no command to rotate the part relative to the axes. It is possible to change the plane in which a sketch is contained by right-clicking the sketch in the FeatureManager and selecting Edit Sketch Plane. However, this may cause errors in subsequent features. Therefore, it is good practice to carefully consider the orientation of the part before selecting the initial sketch plane.

When using either the Centerline Tool or Line Tool, note that there are two different methods for drawing line segments. If you click once to set the first endpoint, move the cursor and then click again to set the other endpoint, you can then move the cursor to create another line segment beginning at the previous endpoint. This technique allows you to create continuous line segments. Double-clicking after the last endpoint has been placed allows you to start a new group of segments, while pressing the Esc key turns off the Line or Centerline Tool. The second method is used to create discontinuous line segments. Click to place the first endpoint, and hold the mouse button while dragging to the other end. Releasing the mouse button places the second endpoint. The cursor can then be moved to the starting point of a new segment.

Select the Line Tool, and draw the two horizontal lines and two diagonal lines shown to form the closed shape shown in Figure 1.177, making sure that the upper corners of the shape are coincident with the upper edge of the solid part.

FIGURE 1.177

Coincident

Select the two diagonal lines and the vertical centerline, using the Ctrl key to make multiple selections, as shown in Figure 1.178. Add a symmetric relation.

FIGURE 1.178

Select the Smart Dimension Tool. Click on both of the diagonal lines to create an angular dimension. Set its value to 40 degrees, as shown in Figure 1.179. Add the two linear dimensions shown in Figure 1.179.

FIGURE 1.179

The sketch should be fully defined.

From the Features group of the CommandManager, select the Revolved Cut Tool, as shown in Figure 1.180. Click on the horizontal centerline to select it as the axis of revolution, as shown in Figure 1.181. Click the check mark to complete the cut, which is shown in Figure 1.182.

FIGURE 1.180

FIGURE 1.181

FIGURE 1.182

Select the Fillet Tool, as shown in Figure 1.183. In the Property-Manager, set the fillet radius to 0.25 inches. Switch to the Trimetric View, and select the two edges shown in Figure 1.184 (shown with the Full preview option selected in the PropertyManager; it may be easier to select the edges with No preview selected.) Select the Rotate View Tool, shown in Figure 1.185.

FIGURE 1.183

FIGURE 1.184

FIGURE 1.185

Rotate the view until the back side of the pulley is visible, then press the Esc key to turn off the Rotate View Tool. Click on the two edges on the back side to be filleted, as shown in Figure 1.186. Click the check mark to add the fillets. Turn off the display of the origin from the Hide/Show Items Tool of the Heads-Up Toolbar, and switch to the Trimetric View.

The filleted part is shown in Figure 1.187. We will complete the part by adding the keyway.

FIGURE 1.186

FIGURE 1.187

Select the face shown in Figure 1.188, and select the Normal To view. Select the Corner Rectangle Tool, and create a rectangle similar to the one shown in Figure 1.189. Press the Esc key to turn off the Rectangle Tool and de-select the rectangle.

FIGURE 1.188

FIGURE 1.189

Select the bottom line of the rectangle and the origin, as shown in Figure 1.190. Add a midpoint relation, which will place the midpoint of the selected line at the origin, as shown in Figure 1.191. Add the two dimensions shown in Figure 1.192 to fully define the sketch.

FIGURE 1.190

FIGURE 1.191

FIGURE 1.192

.250

.600

Select the Extruded Cut Tool, as shown in Figure 1.193. In the PropertyManager, set the type of cut to Through All, as shown on Figure 1.194, and click the check mark to complete the cut, which is shown on Figure 1.195.

FIGURE 1.193

FIGURE 1.194

FIGURE 1.195

The geometry of the part is now complete. Often, we will want to know the mass of the part. First, we must define the part's material.

FIGURE 1.196

In the FeatureManager, right-click on Material and select Edit Material, as shown in Figure 1.196. From the list of available materials, select Cast Carbon Steel from the Steel group, as shown in Figure 1.197.

Note the properties for this material are displayed in Figure 1.197. For the calculation of mass, the density, 0.281793 pounds-mass per cubic inch, is used. The other properties listed are used in stress and thermal analyses.

FIGURE 1.197

Click Apply to accept the material choice, and close the Material box. From the main menu, select Tools: Mass Properties, as shown in Figure 1.198.

FIGURE 1.198

Mass properties are displayed, as shown in Figure 1.199. Note the mass of 8.26 pounds (more precisely, the mass is 8.26 pounds-mass and the weight is 8.26 pounds); the volume and surface area are also calculated. The part's center of mass is indicated on the part, and its coordinates are listed in the Mass Properties window. For this part, the center of mass is at the origin. We know that this is not exactly true; the keyway causes the part to

FIGURE 1.199

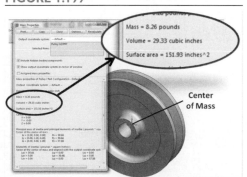

Mass = 8.26 pounds

Volume = 29.33 cubic inches

Surface area = 151.93 inches^2

Center of Mass

DESIGN INTENT | Keeping It Simple

In this exercise, we modeled the pulley using four features: A revolved base, a revolved cut for the V-belt groove, fillets, and an extruded cut for the keyway. These first three features could have been combined by sketching the completed cross-section as a single sketch, as shown here (the Sketch Fillet Tool can be used to create the fillet profiles). Which method is better? Although the finished parts will be identical, breaking a part's creation into simple steps is usually the best approach. Many of the errors that are encountered when making parts are related to sketches—duplicate entities, open contours, endpoints of lines and arcs that do not meet, and under defined geometry are examples of the types of errors that are associated with sketches. The more complex the sketch, the more likely these errors are to occur, and the harder they are to troubleshoot. Also, design changes are easier to make with a part created from many simple steps.

FUTURE STUDY

Dynamics (Kinetics)

In physics, you learned Newton's Second Law:

$$F = ma$$

or the forces (F) acting on a body equal the mass of the body (m) times its acceleration (a). For mechanical engineers designing machines with moving components, application of this law results in the forces required to move a body in a particular manner. In the field of engineering materials, the development of lightweight, strong materials allows moving components to be moved at extremely high accelerations, resulting in higher performance.

The form of Newton's Second Law written above applies to bodies moving in linear or translational motions. Most machine components also move in rotational motions. Newton's Second Law for rotational motions is:

$$T = I\alpha$$

or the torques (T) acting on a body equal the mass moment of inertia (I) times the angular acceleration (α) of the body. This mass moment of inertia is a function of the body's mass and the way in which the mass is distributed relative to the axis of rotation. For example, consider the two wheels shown here. The two wheels are the same diameter and have the same weight, but the wheel on the left has a mass moment of inertia almost twice that of the wheel on the right. If the

wheels were mounted onto shafts, it would take twice as much torque to bring the wheel on the left up to speed as it would the wheel on the right (over the same period of time). While a low mass moment of inertia is usually desirable, a notable exception is a flywheel. The inertia of a flywheel in an engine rotating at a high speed makes it difficult to slow down. As a result, the rotational speed of the engine remains smooth despite small fluctuations in power.

The calculation of mass and mass moments of inertia for complex shapes can be tedious. Therefore, the use of solid modeling software to perform these calculations saves time for engineers performing these types of analysis.

be slightly asymmetric. However, the amount of material removed for the keyway is very small in comparison with the rest of the part, and so the coordinates of the center of mass are zero to the two decimal places shown. Also displayed are the part's moments of inertia, which are important in dynamic analysis and are often very cumbersome to calculate by hand (see the Future Study box).

Close the Mass Properties window, and save the part file.

Often, we will be interested in seeing how changes we make to a part's geometry will affect its mass. A *sensor* allows mass changes to be displayed continuously.

Right-click Sensors in the FeatureManager, and select Add Sensor, as shown in Figure 1.200.

FIGURE 1.200

Note that while sensors are displayed in the FeatureManager by default, the display of sensors and other FeatureManager items (Annotations, Lights, Cameras, and Scenes, etc.) can be modified by right-clicking within the FeatureManager and choosing Hide/Show Tree Items.

Set the type of sensor to Mass Properties: Mass, as shown in Figure 1.201, and click the check mark.

FIGURE 1.201

The mass of the part will now be displayed in the FeatureManager, as shown in Figure 1.202. Note that sensors may be used to track other types of data, such as stress analysis and motion analysis results. Within the Mass Properties group, the volume, surface area, and coordinates of the center of mass can be tracked with sensors.

FIGURE 1.202

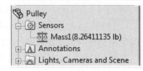

Double-click the V-belt Groove feature (Cut-Revolve1) in the FeatureManager. Double-click the 0.5-inch depth dimension and change it to 0.4 inches, as shown in Figure 1.203. Rebuild the model.

Note that the mass reported by the sensor has changed to reflect the change to the model, as shown in Figure 1.204.

FIGURE 1.203

FIGURE 1.204

Suppose we want to find the mass for a casting of the pulley without the V-belt groove and the keyway. An easy way to do this is to *suppress* those features from the calculations.

Right-click the V-belt Groove (Cut-Revolve1) and select Suppress, as shown in Figure 1.205. Do the same for the keyway (Extrude1).

FIGURE 1.205

FIGURE 1.206

Note that the suppressed entities are shown in gray, as shown in Figure 1.206, and the weight of the casting is shown as 10.25 pounds. The suppressed entities can be restored to the model by right-clicking and selecting Unsuppress.

Close the part file without saving any of the changes made.

PROBLEMS

P1.1 Create a solid model of the stepped shaft, using Extruded Boss/Base and Fillet features.

FIGURE P1.1A

FIGURE P1.1B

∅.750   ∅1.500   R.250   ∅1.125

R.375

3.000   8.000   4.000

Dimensions are in inches

P1.2 Use an Extruded Boss to create a solid model of a 2-foot long segment of a W8X31 steel wide-flange beam (The designation W8X31 denotes wide-flange beam family, 8 inches nominal depth, weight of 31 pounds per foot.) Set the material type to plain carbon steel and find the weight of the part in pounds. Dimensions shown are inches.

Use Equal relations so that the part cross-section is modeled using only the dimensions shown. Use a fully defined sketch of the cross-section.

FIGURE P1.2A

FIGURE P1.2B

8.00

R.50 4 places

.313

8.00

.438

P1.3   Create a solid model of a 5-inch-diameter sphere. Find the formulas for volume and surface area of a sphere, and compare the values calculated from these formulas with those obtained from the Mass Properties Tool.

To create the sphere, sketch and dimension a circle, and draw a line connecting the top and bottom quadrant points of the circle. Trim away the left half of the circle. Add a vertical centerline from the center of the semi-circle, and revolve the sketch 360 degrees about the centerline. Note: Be sure that you have only one centerline in the sketch. Multiple centerlines cause the Revolve command to fail. Also, the complete circle cannot be revolved, as the sketch to be revolved cannot cross the centerline.

FIGURE P1.3A _____   **FIGURE P1.3B** _____

P1.4   Create a model of the torus shown in **Figure P1.4A**. Dimensions are shown in **Figure P1.4B**. Find the formula for the volume of a torus, and compare the value calculated from the formula with that obtained from the Mass Properties Tool.

FIGURE P1.4A _____   **FIGURE P1.4B** _____

P1.5   Create a solid model of this plastic pipe tee. A tee is used to connect pipes together. The type of tee shown here is used to join pipes with solvent welding. A chemical is applied to the inside of the socket, and the pipe is then forced into the socket. The solvent softens the plastic, and when the solvent dries, a strong, permanent joint is created. The sockets are tapered slightly to allow for a tight fit with the pipe.

Set the material to PVC Rigid and find the weight of the tee. Dimensions shown are inches.

(Answer: Weight = 0.1044 pounds. Click the Options tab in the Mass Properties box and increase the number of decimal places, if necessary.)

FIGURE P1.5A

FIGURE P1.5B

SECTION A-A

Sockets are to be tapered 0.50 degrees

P1.6 Create a solid model of this 2-mm-thick steel bicycle disk brake rotor. Calculate the mass of the part, using AISI 304 stainless steel as the material. Dimensions shown are mm.

Note: Although this part can be modeled with a single extrusion from a complex sketch, you will find it easier to extrude a solid disk and then use a series of simple extruded cuts and circular patterns.

FIGURE P1.6A

FIGURE P1.6B

P1.7   Create a solid model of the pleated filter element. Dimensions shown are inches, and the diameter dimensions shown are nominal dimensions. There are 36 pleats in the part, and the part is 8 inches long.

To create this part, start a sketch with two construction circles representing the nominal inner and outer diameters. Add and dimension two lines representing one pleat, and use a circular pattern to copy these two lines into the other 35 positions. Select the Extruded Boss Tool. Since the sketch is an open contour, a thin-feature extrusion will be created. Set the thickness to 0.02 inches, and set the type as Mid-Plane.

FIGURE P1.7A   **FIGURE P1.7B**

P1.8   The ancient Greek mathematician Pappus of Alexandria theorized that the volume of a solid of revolution could be calculated by multiplying the area of the solid's cross-section by the distance travelled by the centroid (geometric center) of the cross-section during a full revolution. Demonstrate the application of this theorem (refered to as Pappus' Centroid Theorem or the Pappus-Guldinus Theorem) by creating the solid of revolution shown in **Figure P1.8A** as follows:

1. Create the sketch shown in **Figure P1.8B** in the Front Plane. Note that the section is symmetric about the horizontal axis. Use relations as necessary to fully define the sketch with only the dimensions shown. (Note: the sketch relations icons are turned off for clarity in the figure.)

FIGURE P1.8A   **FIGURE P1.8B**

2. Select Tools: Section Properties. Note the area of the section, as well as the x-coordinate of the centroid location (shown as dimension *d* in **Figure 1.8C**).

3. Calculate the volume of the solid of revolution by multiplying the distance traveled by the centroid during the revolution (*d* times 2π) by the area of the section.

4. Use the Revolved Boss/Base Tool to create the solid by revolving the section about the vertical axis. Select Tools: Mass Properties and compare the volume to that calculated in step 3 above.

FIGURE P1.8C

P1.9   Create the body of the flange shown in **Figure 1.1** and detailed in **Figure 1.2** by sketching its cross-section in the Right Plane and revolving it about the flange's axis. Include the fillets in the sketch, using the Sketch Fillet Tool.

P1.10 Create a solid model of the connecting rod shown in **Figure P1.10A**. Use the dimensions shown in **Figure P1.10B**. Assign the material to be Cast Carbon Steel, and add a sensor to determine the mass. Notes: When sketching the shape of the rod, use tangent relations between the diagonal edges and the rounded ends. To maintain the symmetry of the rod, make your sketches in the Front Plane and choose Midplane as the type of extrusion. This will cause the sketch to be extruded an equal distance in both directions from the Front Plane.

FIGURE P1.10A   **FIGURE P1.10B**

P1.11 Consider the connecting rod model created in **P1.10**. Due to weight restrictions on the design, the weight of the connecting rod must be reduced to 1.70 lb. Add a recessed pocket like the one shown in **Figure P1.11** to both sides of the connecting rod (to maintain symmetry), and select appropriate dimensions to reduce the weight of the part to an acceptable level. Notes: Experiment with the Offset Entities Tool when creating the recessed pocket. This tool allows for new entities to be created at a specified distance away from selected lines or arcs. Also, try the Mirror Tool to create the second pocket, with the Front Plane as the mirror plane. These tools will be explained in detail in later chapters.

FIGURE P1.11

CHAPTER 2

Engineering Drawings

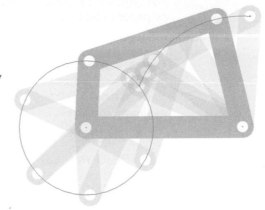

Introduction

Even companies that design parts exclusively with solid modeling software still need to produce 2-D drawings. A multiple-view 2-D drawing is the best way to document the design, showing all of the dimensions necessary to produce the part. Since no manufacturing process can produce "perfect" parts, the tolerances (variations from the stated dimension of a part) allowed for important dimensions are also shown on 2-D drawings.

The SolidWorks program allows for 2-D drawings to be quickly and easily produced from 3-D models. The drawing will be fully associative with the part file, so that changes to the part will be automatically reflected in the drawing, and vice versa.

Chapter Objectives

In this chapter, you will:

- make a 2-D drawing from a SolidWorks part,

- create a custom drawing sheet format, and,

- use eDrawings® software to create a drawing file that allows for easy file sharing and collaborative editing[1].

2.1 Drawing Tutorial

In this tutorial, a 2-D drawing of the flange from Chapter 1 will be made.

Open the SolidWorks program. Click on the New Document Tool, as shown in Figure 2.1. Select the Drawing icon, as shown in Figure 2.2, and click OK.

FIGURE 2.1

FIGURE 2.2

---

[1] eDrawings is a registered trademark of the SolidWorks Corporation, 300 Baker Avenue, Concord, MA 01742.

As if you were making a drawing by hand, your first step is to select the size and type of paper to be used. Later we will look at using sheet formats with a title block, but for now let's use a blank 8-1/2 X 11-inch sheet (A-Landscape).

Choose a paper size of A-Landscape, and click on the box labeled "Display sheet format" to clear the check mark, as shown in Figure 2.3. Click OK to close the dialog box.

FIGURE 2.3

If the Model View box appears in the PropertyManager, as shown in Figure 2.4, click on the X to close the box.

FIGURE 2.4

Before bringing in the flange model data, we will set a few options.

Select the Options Tool, as shown in Figure 2.5. Under the System Options tab, select Drawings: Display Style, and set the hidden line and tangent displays as shown in Figure 2.6.

FIGURE 2.5

Hidden lines are usually displayed in standard drawing orthographic views, but not in section or detail views. Tangent edges are usually not shown in drawing views. Whether or not to show hidden and tangent edges is sometimes dependent on the complexity of the part. As we will see later, we can change these display options for each model view.

FIGURE 2.6

Select Colors: Drawings, Paper Color. Select Edit, as shown in Figure 2.7. Select white for the paper color and click OK. Choose the option labeled "Use specified color for drawings paper color." Selecting this option overrides the default setting of the crumpled paper image as the drawing sheet background.

FIGURE 2.7

Under the **Document Properties tab**, select **Drafting Standard** and set the **dimen-sioning standard to ANSI**, as shown in Figure 2.8. Select **Detailing**, and turn on the automatic display of center marks and center lines for circular features, as shown in Figure 2.9. Select **Units**, and set the **Unit system to IPS (inch, pound, second)**, and the **decimal places to 2**, as shown in Figure 2.10. Set the decimal places for angles to none. Click OK.

FIGURE 2.8

FIGURE 2.9

FIGURE 2.10

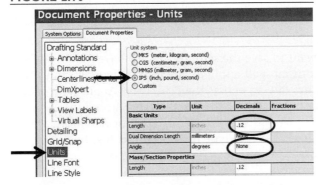

Although the part was modeled with the decimal places set to 3, the number of decimals does not affect the accuracy of the model, only the way dimensions are displayed. For a drawing, the number of decimal places should be related to the tolerance level of most of the dimensions. Note that there is an option for setting the units of dual dimension lengths. Often, drawings show one set of dimensions in the primary unit system and a second set of dimensions in brackets. In this way, both US and SI units can be displayed on a single drawing. Since we will not be using dual dimensions, we do not need to change the default setting.

FIGURE 2.11

Before adding our drawing views, we will set the type of projection to third angle. The third-angle option displays the Front, Top, and Right Views, with the Front View in the lower left position (standard for most US drawings), while the first-angle option displays the Front, Top, and Left Views, with the Front View in the upper left position (standard for most European drawings). For example, consider the part shown in **Figure 2.11**. The first- and third-angle projections of this part are shown in **Figure 2.12**.

In the drawing space, right-click and select Properties from the menu that appears. Set the Type of projection to Third angle, as shown in Figure 2.13, and click OK to close the Sheet Properties box.

FIGURE 2.12

FIGURE 2.13

Before proceeding, we may want to save the settings that we have made. The settings under the System Options tab will apply to any future drawings, but those under the Document Properties tab, as well as the First/Third Angle Setting, will revert to their previous values unless saved. These settings are stored in a *template* file. When we start a new drawing and select Part, Drawing, or Assembly, we are opening the template associated with that type of SolidWorks file. In Chapter 1, we modified the Part template. We will now perform a similar operation to modify the Drawing template, saving our changes for future drawings.

Select File: Save as from the menu. From the Save as type pull-down menu, select Drawing Templates, as shown in Figure 2.14. The default template name, Drawing.drwdot, will be displayed in the file list, as shown in

FIGURE 2.14

Figure 2.15. Click on the template name to select it, and click Save. When asked if you want to overwrite the existing file, click Yes.

FIGURE 2.15

The settings that you specified will be applied when you open a new drawing. One note of caution: as with the Part template in Chapter 1, the next time you save a file, the default folder will be the one where the templates are stored. Be sure to change the folder to the one where you want to store the file.

When we modeled a part in Chapter 1, we found most of the commands we needed in the CommandManager. In the drawing mode, the CommandManager has three groups of commands by default: View Layout, Annotation, and Sketch. The group is changed by clicking on the appropriate tab below the CommandManager.

We are now ready to import the geometry from the part file.

From the View Layout group of the CommandManager, select the Model View Tool, as shown in Figure 2.16.

FIGURE 2.16

If you have the flange open in another window, then it will appear in the PropertyManager. If not, then click the Browse . . . button and find the flange part file where you stored it, as shown in Figure 2.17. In the Property-Manager, select Multiple Views and click on the Front and Top Views to select them, as shown in Figure 2.18. Click the check mark to place the views.

Note the "A" designation shown of the Front and Top Views in Figure 2.18. This indicates that model dimensions are associated with these views. At a minimum, all views with associated dimensions should be brought into the drawing.

As you pass the cursor over each drawing view, notice that a rectangular box that defines the boundary of that view appears. Moving the cursor over the drawing view boundary displays the move arrows, as

FIGURE 2.17   FIGURE 2.18

FIGURE 2.19

shown in Figure 2.19. **When the move arrows are displayed, you can click and drag the drawing view to a new location. Note that the views remain in alignment as they are moved.**

Note that we need only the Front and Top Views for this part, since the Right View shows no information that cannot be seen in the Top View. Also, note that the drawing scale was automatically selected. You can override the scale by right-clicking in the drawing area, selecting properties, and defining a new scale.

Now we will import the dimensions that were used to create the model.

Click the Annotation tab of the CommandManager, and select the Model Items Tool as shown in Figure 2.20. In the PropertyManager, select Entire model as the source. Also check the boxes labeled "Import items into all views" and "Elimate duplicates," as shown in Figure 2.21. Click the check mark to import the model items (dimensions) into the drawing.

FIGURE 2.20

While the PropertyManager for Model Items contains options for controlling the import of dimension, notes, tolerances, etc., accepting the default options is usually sufficient.

You can click and drag on the value of any dimension to reposition it within its associated drawing view.

Click and drag the dimensions on the drawing sheet so that they can all be clearly seen, as shown in Figure 2.22.

FIGURE 2.21

FIGURE 2.22

You can also change how the arrows are displayed on linear dimensions, such as for the 2.25-inch dimension shown in **Figure 2.23**. When you click once on the dimension value to select it, the dimension will appear in green and dots will appear on the arrowheads. When you click on one of these dots, the arrowheads will switch from outside to inside or vice versa, as shown in **Figure 2.24**.

FIGURE 2.23

FIGURE 2.24

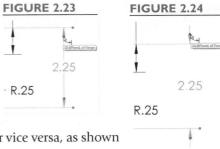

Dimensions are shown in the view in which they were created in the part. For example, the 5.5-inch and 2.75-inch diameters are shown in the Front View, since they were created from circles sketched in the Front Plane (or sketch planes parallel to the Front Plane). However, standard drawing practice is for the diameter of a solid cylindrical feature to be shown in a view normal to the one in which it appears as a circle. Therefore, we want these dimensions in the Top View. Rather than deleting these dimensions from the Front View and adding them to the Top View, we will simply move them.

Press and hold down the Shift key. Click and drag the 5.5-inch dimension until the cursor is within the boundaries of the Top View. Release the mouse button, and the dimension will be placed in the Top View, and shown on Figure 2.25. (Make sure to release the mouse button before releasing the Shift key; otherwise, the move operation will not work.) Repeat for the 2.75-inch dimension.

FIGURE 2.25

We can modify the dimension text to include notes or additional information. For example, since the bolt holes created with the Circular Pattern Tool are all of the same diameter, we can augment the 0.50-inch-diameter text to include the fact that this dimension is typical of all four holes.

Click on the 0.50-diameter value to select it. The Dimension PropertyManager will open. In the PropertyManager, add the text "4X" in the Dimension Text box, as shown in Figure 2.26. (Note that the symbols that already appear in the box are the code for the diameter symbol and the link to the part dimension. We will not modify these.) Click the check mark to apply the change.

FIGURE 2.26

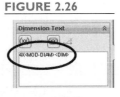

Holes diameters should be shown with leaders, rather than as linear dimensions.

DESIGN INTENT | Exploiting Associativity

In Chapter 1, we defined the position of the bolt holes in the flange by creating a construction circle and specifying the diameter dimension of the circle. Alternatively, we could have located a single hole by specifying its radial location from the center point of the flange. As we modeled the flange, either dimensioning technique would properly locate the first hole in our hole pattern, and it would seem that either alternative would be acceptable. When dimensioning the drawing, however, we see that drawing a bolt circle and dimensioning its diameter is the preferred way to show the radial position of the bolt holes. If the distance were defined by a radial dimension, then we could simply delete the radial dimension in the drawing, add a construction circle, and add a new dimension. However, the new dimension would not be associative with the part. That is, changing the added 4.25-inch dimension would not change the bolt hole positions, and changing the radial dimension in the part would not cause the drawing dimension to be updated. Therefore, maintaining full associativity between the part and the drawing would require editing of the part file.

FIGURE 2.27

Right-click on the 1.5-inch diameter dimension of the center hole, and choose Display Options: Display as Diameter, as shown in Figure 2.27. Click and drag the dimension to a convenient location, as shown in Figure 2.28.

The options for changing the appearance of dimensions, dimension lines, and leaders are numerous, and we will cover only a few in this text. By right-clicking on any dimension and selecting Display Options or Properties, you will find that it is easy to change the appearance of the dimension.

FIGURE 2.28

The font for any dimension can be changed individually, but if you want to change the font type or size for all dimensions, you can do that from the Document Properties menu.

Select the Options Tool. Under Document Properties, select Dimensions, click the Font button, and select a font and font size. The font size may be input in inches or points. Click OK to apply the selected font.

A section view will help show the center hole and the chamfer more clearly.

FIGURE 2.29

Click the View Layout tab on the Command-Manager, and select the Section View Tool as shown in Figure 2.29.

The line tool becomes active when the Section View Tool is selected, as shown by the line icon next to the cursor. Since we want to draw our line exactly through the center of the part, we need to "wake up" a center mark and quadrant points for placement of the line's endpoints.

FIGURE 2.30

Without clicking any mouse button, move the cursor over the edge of one of the circular features, as shown in Figure 2.30, and hold it there momentarily. Move the cursor outside of the edge of the flange.

FIGURE 2.31

When the dotted line appears that indicates alignment with a quadrant point, as in Figure 2.31, click and drag a horizontal line completely through the part, as shown in Figure 2.32 (make sure that the horizontal icon appears by the line before clicking).

FIGURE 2.32

When you finish the line, a section view is created. Move the cursor to place the section view. By default, the view will be kept in alignment with the Front View. Drag the section view away from the Front View, and click once to place the view, as shown in Figure 2.33.

FIGURE 2.33

Right-click within the section view boundary box, and from the menu that appears, select Alignment: Break Alignment, as shown in Figure 2.34. Move the cursor over the edge of the section view boundary box so that the move arrows appear. Click and drag the view to a new location on the sheet, as shown in Figure 2.35.

FIGURE 2.34

FIGURE 2.35

When we created the section view, the direction of the view relative to the section line was down, toward the bottom of the sheet. This resulted in a section view that is upside-down from our preferred orientation of the part. This is easily modified.

FIGURE 2.36

Double-click on the section line in the Front View. The arrows defining the direction of the section will be reversed, as shown in Figure 2.36.

The section view will not be immediately reversed. Instead, it appears crosshatched, indicating that it needs to be rebuilt.

Click the Rebuild Tool, as shown in Figure 2.37.

FIGURE 2.37

Rebuild (Ctrl+B)
Rebuilds the part, assembly, or drawing.

The section view appearance is now consistent with the direction of the section arrows, as shown in Figure 2.38.

FIGURE 2.38

SECTION A-A

Section views are usually shown without the hidden lines displayed.

Select the section view, and click on the Wireframe with Hidden Lines Removed Tool from the Heads-Up View Toolbar, as shown in Figure 2.39.

FIGURE 2.39

Hidden Lines Removed
Displays only those model edges that can be seen from the current view orientation.

FIGURE 2.40

SECTION A-A

The section view should now appear as in Figure 2.40.

The chamfer is difficult to see at the default scale (1:2). A detail view that enlarges the chamfer region will be helpful.

Select the Detail View Tool from the View Layout group of the CommandManager, as shown in Figure 2.41.

The Circle Tool will be activated automatically. Drag out a circle around the area to be included in the detail view, as shown in Figure 2.42. Move the cursor to the position where you want the detail view to appear, and click to place the view, as shown in Figure 2.43. If a centerline appears in the detail view, select it and press the Delete key.

FIGURE 2.41

Detail View
Adds a detail view to show a portion of a view, usually at an enlarged scale.

FIGURE 2.42

R = 0.52
View

FIGURE 2.43

DETAIL B
SCALE 1 : 1

FIGURE 2.44

Note that the detail view has a scale of 1:1. The scale of any drawing view can be changed, and we will change the scale of the detail view to enlarge the chamfer area further.

With the detail view selected, check the "Use custom scale" box in the PropertyManager, and change the scale to 2:1, as shown in Figure 2.44. Click the check mark to apply the scale.

Dimensions can be imported into section and detail views. In this case, the 45-degree angle of the chamfer can be seen best in the detail view.

Click in the white space of the detail view boundary box to select this view, and select the Model Items Tool from the Annotation group of the CommandManager, as shown in Figure 2.45. Click the check mark to import dimensions into the view. The 45-degree dimension will be added, as shown in Figure 2.46. Press Esc to deselect the detail view.

FIGURE 2.45

FIGURE 2.46

DETAIL B
SCALE 2 : 1

We would also like to show the 0.080-inch width of the chamfer on the detail view, but that dimension has already been imported into the Front View. We will move the dimension from the Front View to the detail view.

Press and hold the Shift key. Click and drag the 0.080-inch chamfer dimension from the Front View into the detail view boundary box. When you release the mouse button, the dimension will be moved into the detail view. Release the Shift key, and drag the dimension to its final position, as shown in Figure 2.47.

Change the font of the labels on the section and detail views by clicking on each text group, clearing the check from the "Use document's font" box in the PropertyManager, and clicking on the Font button, as shown in Figure 2.48. Increase the size of the font.

FIGURE 2.47

DETAIL B
SCALE 2 : 1

FIGURE 2.48

Your drawing should now appear as in **Figure 2.49**.

As we noted earlier, the number of decimal places displayed in the drawing will relate to the tolerances applied to that dimension. Often, certain dimensions are more critical than others and will need to have tolerance values specified. For example, suppose that the center 1.5-inch diameter hole needs to have a tight tolerance to allow for a good fit with the part that will be inserted into the hole. When that dimension is selected, a number of options appear in the PropertyManager.

FIGURE 2.49

Click on the number portion of the 1.50-inch dimension. In the PropertyManager, shown in Figure 2.50, select a tolerance type of Bilateral from the drop-down menu. Set both the dimension and tolerance values to display three decimal places (.123). Enter a value of 0.003 as the plus tolerance. Leave the minus tolerance as zero. Click the check mark.

The dimension should now appear as in **Figure 2.51**, indicating that the diameter can be no smaller than the 1.500-inch nominal dimension, but can be up to 0.003 inches larger.

FIGURE 2.50

FIGURE 2.51

$$\varnothing\,1.500^{+.003}_{-.000}$$

In addition to dimensions with tolerances, *geometric tolerances* are often added to drawings in order to fully define the acceptable limits of a part's geometry. For example, suppose that the back surface of the flange is to mate to another part with a gasket between the two parts to create a seal. If the back of the flange is warped, it may not allow for a proper seal, even if all dimensional tolerances are met. To ensure a good surface for sealing, we might need to add a flatness specification.

FUTURE STUDY

Manufacturing Processes, Geometric Dimensioning and Tolerancing, and Metrology

In this example, we placed a relatively tight tolerance on the diameter of the center hole so that a tight fit could be realized between the flange and the part that fits into the hole. How do you know what the tolerance should be? Perhaps a better question is: what tolerance is required for the part to function as designed in an assembly? The answer to this question will in many cases dictate the manufacturing process that can be used. Many engineering students study manufacturing processes, but it is impossible to learn the specific details of all of the manufacturing processes in use today. On the job, engineers should learn as much as possible about the manufacturing processes used by their company and its suppliers, in order to understand what tolerances are practical and economical.

Applying dimensions and tolerances to engineering drawings is a more complex topic than it first appears. For example, consider a simple part such as a pin that is required to fit into a hole. If we simply define the diameter of the pin and place tolerances on the diameter, are we sure that the pin will fit into the hole? What if the pin is bent slightly? Its diameter might be within the limits defined by the tolerance, but it may not work in its intended purpose. The process of Geometric Dimensioning and Tolerancing (GD&T) allows a designer to specify the acceptable condition of a part, considering its function. In the example we just discussed, a straightness tolerance might be required. In the drawing tutorial in this chapter, we added a flatness specification to a surface that was important for sealing. Proper application of GD&T standards can actually reduce the cost of making many parts, since they allow control of the important features of a part more efficiently than simply tightening the tolerance values of all dimensions. At many companies, the detailing of drawings is performed by a drafting department, and engineers document designs with less-formal drawings and sketches. At many smaller companies, however, engineers detail their own drawings. For these engineers, study of GD&T standards and practices is necessary.

A related area of study is metrology, the science of measurements. Many features are difficult to measure accurately with traditional methods, and computer-controlled Coordinate Measuring Machines (CMMs) are now widely used in industry. Using statistical methods with measured data, quality assurance engineers track variations and work to control the processes that are used to produce the components. This method of Statistical Process Control (SPC) allows problems to be detected and corrected before defective parts are produced.

FIGURE 2.52

In the Top View, click on the line corresponding to the back surface of the flange, as shown in Figure 2.52. Select the Geometric Tolerance Tool from the Annotation group of the CommandManager, as shown in Figure 2.53.

FIGURE 2.53

FIGURE 2.54

FIGURE 2.55

In the dialog box, select the flatness symbol from the pull-down symbol menu, as shown in Figure 2.54. Enter 0.010 as the value for Tolerance 1, and a preview of the tolerance call-out will appear, as shown in Figure 2.55. Click to place the annotation in the desired location. Click OK to apply the tolerance, and drag the call-out box to the desired location, as shown in Figure 2.56.

Since we are not using a title block with default tolerances shown, we will specify them in a note.

FIGURE 2.56

FIGURE 2.57

Select the Note Tool from the Annotation group of the CommandManager, as shown in Figure 2.57. Drag the cursor to the approximate location where the note will be placed (near the bottom of the drawing), and click. Choose a font type and size from the toolbar that appears. In the text box that appears at the location where you clicked, begin typing the text of the notes shown in Figure 2.58.

Use the Enter key to start a new line. The ± symbol is inserted by clicking on the Add Symbol Tool in the PropertyManager (Figure 2.59) and selecting Plus/Minus from the list of available symbols, as shown in Figure 2.60. Click outside of the text box to place the note. Press the Esc key to turn off the Note Tool; otherwise, subsequent mouse clicks will place duplicate notes on the drawing.

If desired, you can change the font type and size by double-clicking the note and changing the font parameters on the toolbar that appears.

FIGURE 2.58

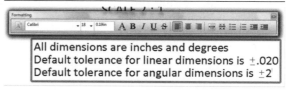

All dimensions are inches and degrees
Default tolerance for linear dimensions is ±.020
Default tolerance for angular dimensions is ±2

FIGURE 2.59

FIGURE 2.60

Since we defined all of the fillets with a single fillet command in the part, only one of the fillets is dimensioned in the drawing. We might prefer to call out the fillet radius in a note, yet maintain associativity with the part.

Double-click on the notes. Move the cursor to the end of the last line and press the Enter key to move to a new line. Type in "All fillets and rounds are", and click on the .25-inch dimension, as shown in Figure 2.61. **The dimension's value will appear in the text box. Click outside of the text box to end editing of the note. The completed note is shown in** Figure 2.62.

FIGURE 2.61

FIGURE 2.62

All dimensions are inches and degrees
Default tolerance for linear dimensions is $\pm.020$
Default tolerance for angular dimensions is ± 2
All fillets and rounds are R.25

FIGURE 2.63

If you change the value of the radius, then the value in the note will change accordingly. Now that the fillet radius is shown in a note, displaying it on the drawing is unnecessary. However, deleting the dimension will produce an error, since the note links to the value of the dimension. Therefore, we will hide the dimension.

Right-click on the R.25 dimension and select Hide from the menu, as shown in Figure 2.63.

If you want to show the dimension later, you can do so by selecting Hide/Show Annotations from the main menu. Any hidden dimensions will show up light, as shown in **Figure 2.64**, and clicking on them will change them back to visible dimensions. Similarly, clicking on visible dimensions will hide them. This is a good method for hiding several dimensions at once. When finished, select Hide/Show Annotations again or press the Esc key to return to the normal editing mode.

FIGURE 2.64

A pictorial view such as a trimetric view can often be helpful in interpreting a 2-D drawing.

FIGURE 2.65

Select the Model View Tool from the View Layout group of the CommandManager, as shown in Figure 2.65. **The part can be selected by double-clicking the part name in the PropertyManager. In the PropertyManager, select the Trimetric View as the orientation of the new model view, and the shaded with edges mode as the display style, as shown in** Figure 2.66. **Move the cursor to the desired location of the new view, and click to place the view.**

FIGURE 2.66

Right-click on the Trimetric View, and select Tangent Edge: Tangent Edges with Font, as shown in Figure 2.67.

Move the drawing views and notes around the sheet so that they can all be seen clearly. Add a note identifying the scale as 1:2. To see how the drawing will look when printed, select Print Preview from the File menu.

Your drawing should appear similar to Figure 2.68.

FIGURE 2.67

FIGURE 2.68

Select File: Save and save the drawing as "Flange," with a file type of .slddrw (SolidWorks Drawing).

Open the Flange part file, and experiment with changing dimensions and noticing the updating of the drawing. Similarly, change some drawing dimensions and observe the changes in the part.

A tool which can be helpful in visualizing a drawing view is the 3-D Drawing View Tool, which is found on the Heads-Up View Toolbar, as shown in Figure 2.69. With any of the standard drawing views or the section view selected, selecting the 3-D Drawing View tool allows you to rotate that view within its boundaries, as shown in Figure 2.70.

FIGURE 2.69

FIGURE 2.70

Clicking the check mark in the pop-up toolbar next to the drawing view turns the 3-D Drawing View Tool off and resets the view to its 2-D orientation.

2.2 Creating a Drawing Sheet Format

Most engineering drawings include a title block and a border. The title block can include a large amount of information. In addition to basic data, such as the name of the part, the part number, the scale, the date created, etc., many companies require the names or initials of reviewers who must approve the drawing before it is released to users. Some require a revision history that tracks changes made to the drawing. Others require the listing of information about where the part described on the drawing is used in assemblies.

The SolidWorks program includes several default title blocks and borders, called *sheet formats*. These sheet formats can be edited to fit the particular needs of a company. The sheet formats are different for every standard paper size. In this section, we will create a simple sheet format for an A-size drawing. Our format will include some basic information without taking up much room on the sheet. Rather than modifying an existing title block, we will build ours from scratch.

Open a new drawing. Choose A-Landscape as the paper size; leave the "Display sheet format" box unchecked, as shown in Figure 2.71**. If the Model View command starts automatically, check the X in the PropertyManager to close it, as shown in** Figure 2.72**.**

FIGURE 2.71

Note that you can prevent the Model View command from starting automatically by clearing the check mark from the box labeled "Start command when creating new drawing" in the PropertyManager. However, since we will usually be creating a drawing from an existing part, we will allow the command to continue to start automatically.

FIGURE 2.72

FIGURE 2.73

Right-click in the drawing space. Select Edit Sheet Format, as shown in Figure 2.73**.**

In a drawing, you can toggle between editing the drawing itself and editing the sheet format, which contains the title block and border.

Select the Corner Rectangle Tool from the Sketch group of the CommandManager, as shown in Figure 2.74. Drag out a rectangle that fills most of the page, as shown in Figure 2.75.

FIGURE 2.74

FIGURE 2.75

Right-click on the menu bar or the CommandManager, and select Line Format from the list of available toolbars, as shown in Figure 2.76. With the rectangle still selected, choose a heavier line weight from the Line Format Toolbar, as shown in Figure 2.77.

FIGURE 2.76

FIGURE 2.77

Click on the lower left corner of the rectangle to select it, as shown in Figure 2.78. In the PropertyManager, input its x and y coordinates as 0.375 inches each, as shown in Figure 2.79. Click on the Fix icon.

FIGURE 2.78

FIGURE 2.79

Note that if your default units are mm, you need to change to the ips system.

Select the upper right corner of the rectangle. Set its coordinates to x = 10.625 inches and y = 8.125 inches. Fix this point.

FIGURE 2.80

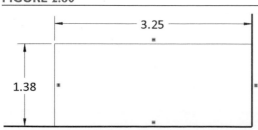

In the lower right corner of the sheet, drag out a rectangle from the corner of the border. Select the Smart Dimension tool, and dimension the rectangle as 3.25 inches by 1.375 inches, as shown in Figure 2.80.

FIGURE 2.81

Select the Line Tool, and add the horizontal and vertical lines shown in **Figure 2.81**. Be sure that the endpoints snap to lines and not specific points (such as midpoints).

Select the Smart Dimension Tool, and add the 1-inch and 1.375-inch dimensions shown.

FIGURE 2.82

Select the Line Tool, and drag out a horizontal line from the midpoint of the vertical line shown in **Figure 2.82** to the vertical line on the right side of the title block.

Add the other three lines shown in **Figure 2.83**, and add dimensions and relations as shown so that the drawing is fully defined.

FIGURE 2.83

We will now hide the dimensions just added.

From the main menu, select View: Hide/Show Annotations. Click on each dimension, and it will turn gray, as shown in **Figure 2.84**. When all dimensions are selected, press the Esc key to hide them.

FIGURE 2.84

You do not need to turn off the display of the sketch relations; they will not appear when we toggle back from editing the sheet format to editing the drawing.

Select the Note Tool from the Annotation group of the CommandManager. Click in any blank area to place the note, and set the font as desired and the size to 10 point. Type in the text as shown in Figure 2.85, using the Enter key to move to a new line. Choose the Center tool to center the text within the box. Click outside of the text box, and press the Esc key to end the Note command (if you do not press the Esc key, then you can click in different areas of the drawing to place the same note in multiple locations). Click and drag the note to the position shown in Figure 2.86.

FIGURE 2.85

FIGURE 2.86

Add the rest of the text shown in Figure 2.87, adding your school/company name and your name to the title block. Use the font sizes shown in Figure 2.87. When opening a new note, select the font size before typing in the text. Font sizes can be easily changed by double-clicking each note and selecting a new font size.

FIGURE 2.87

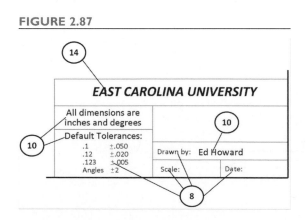

Note that it is sometimes difficult to move a text entity to the position desired because of auto-aligning features. To have more precise control when locating a note, press and hold the Alt key while moving the text. This temporarily suspends the auto-aligning.

We will now add the scale value to the drawing, linking it to the drawing sheet's scale.

Open a new note. In the PropertyManager, select the Link to Property Icon, as shown in Figure 2.88. From the pull-down menu in the dialog box, select "SW-Sheet Scale," as shown in Figure 2.89, and click OK.

FIGURE 2.88

FIGURE 2.89

The sheet's scale will now be added to the note. Move the scale note into the correct location in the title block, as shown in Figure 2.90. If the scale of the drawing is changed, the title block will update automatically.

Click outside the text box, and press the Esc key to end the Note command. Right-click in the drawing area and select Edit Sheet, as shown in Figure 2.91.

FIGURE 2.90

FIGURE 2.91

The title block and border are now in the background, and cannot be edited without toggling back to Edit Sheet Format.

From the main menu, select File: Save Sheet Format, as shown in Figure 2.92. Give the sheet a unique name and save it to the default directory, as shown in Figure 2.93. Close the drawing. Select Don't Save when asked if you want to save the drawing file.

FIGURE 2.92

FIGURE 2.93

The drawing sheet format just created can be selected when beginning a new drawing, and can be applied to existing drawings. We will now apply the format to the flange drawing.

Open the flange drawing created earlier. Right-click in the drawing space (away from any drawing view) and select Properties. In the Sheet Properties box, select the sheet format just created from the pull-down list, as shown in Figure 2.94. Check the box labeled "Display sheet format" and click OK.

FIGURE 2.94

Move the drawing views as necessary. Delete the notes concerning units and tolerances, which are now covered in the title block. Add notes showing the drawing title and the date. The finished drawing is shown in Figure 2.95. **Save the drawing for use in the eDrawings tutorial that follows.**

FIGURE 2.95

2.3 Creating an eDrawing

One disadvantage of using drawings that are fully associative with part files is that in order to share the file with other users, they must have access to both the drawing file and the associated part file. One option for sharing drawings without transferring the part file is to create a *detached drawing*. When you open a detached drawing, it is not necessary to have the model file loaded. In addition to viewing and printing the drawing, annotations can be added to the drawing. When the model is required, for example if you attempt to edit a dimension imported from the model, then you are prompted to load the model from which the drawing was created. To save a drawing as a detached drawing, choose File: Save As from the main menu and select a file type as "Detached Drawing".

Another option for sharing drawings with others, including those who do not have access to SolidWorks software, is to create an eDrawing. The eDrawings Viewer can be downloaded for free, and the eDrawings Publisher also allows the creation of executable files that can be read on any computer. Even within organizations using SolidWorks software, eDrawings are extensively used to share information. Markup tools allow users to make comments and corrections, and the small file sizes of eDrawings make exchange of data via e-mail more practical.

To save a SolidWorks drawing as an eDrawing, simply select File: Save As from the main menu and select eDrawing (.edrw) as the file type. The eDrawing file format is very efficient, and file sizes produced are small. For example, the file sizes of the part and drawing files of the flange in Chapters 1 and 2 are:

- Part file: 281 kB
- SolidWorks drawing file: 346 kB
- eDrawings file: 76 kB

The eDrawings file is smaller than most image files (.bmp, .pdf, etc.) that could be used to electronically communicate a drawing, but as we will see, an eDrawing is more than a 2-D image. The viewer can rotate any of the drawing views in 3-D space.

To create an executable file or to mark up an eDrawing, it is necessary to have the eDrawings program installed on your computer.

Open the flange drawing file that you saved in the previous section. From the main menu, select Tools: Add-Ins. If the box next to eDrawings is not checked, click on the box to activate eDrawings, as shown in Figure 2.96. **By checking the "Start Up" box, you can have the eDrawings Add-In loaded whenever you start a new SolidWorks session. Click OK.**

Select File: Publish eDrawings from the main menu, as shown in Figure 2.97. **An eDrawing is created and the eDrawings Publisher is opened, as shown in** Figure 2.98.

FIGURE 2.96

FIGURE 2.97

FIGURE 2.98

Note: Once you have installed the eDrawings Publisher to your computer, you can open the program from the Start menu on your computer or an icon on your desktop. Installing eDrawings as a SolidWorks add-in simply provides easy access to the program from the SolidWorks environment.

By default, the eDrawing views are displayed in the shaded mode. The display mode can easily be changed to show line drawings.

FIGURE 2.99

Click on the Shaded Tool, as shown in Figure 2.99.

The eDrawings Viewer allows drawings to be animated. The Play Tool automatically switches from one view to another. The Next Tool allows you to control the animation steps.

Click the Next Tool (Figure 2.100) repeatedly to switch from one view to another, including section and detail views.

FIGURE 2.100

When the Section View is active, select the Rotate Tool, as shown in Figure 2.101. Click and drag to rotate the model view, as shown in Figure 2.102.

Select the Home Tool, as shown in Figure 2.103 to return to an overview of the drawing.

FIGURE 2.101

FIGURE 2.102

FIGURE 2.103

An important feature of eDrawings is that comments can be added by reviewers. To illustrate, we will add a comment for draft (taper to allow the part to be easily removed from a mold) to be added to the outer surface of the 2.75-inch diameter boss.

Select the Markup Tool from the left side of the screen, as shown in Figure 2.104.

FIGURE 2.104

FIGURE 2.105

Select the Options button, as shown in Figure 2.105.

In the box that appears, you can set preferences for your comments. Several reviewers can add comments to the same drawing, and each reviewer can have a unique color.

Enter a name for the reviewer, and select a color and font type/size, as shown in Figure 2.106. Click OK.

FIGURE 2.106

From the markup tools, select the Cloud with Leader Tool, as shown in Figure 2.107. Click on the edge to be drafted, as shown in Figure 2.108. In the pop-up box that appears, type in the text of the comment, as shown in Figure 2.109. Click the check mark to close the box, and then click at the location where the comment is to be placed. The comment as it appears on the drawing is shown in Figure 2.110.

FIGURE 2.107

FIGURE 2.108

FIGURE 2.109

FIGURE 2.110

Comments are listed on the left side of the screen, as shown in Figure 2.111, allowing the person making the corrections to the drawing to see who made each comment. In this way, eDrawings allow efficient collaboration among different groups working on a project.

FIGURE 2.111

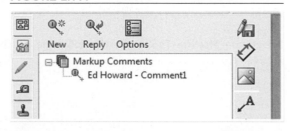

We can also "Stamp" a drawing, much as paper drawings can be stamped to limit its distribution, identify it as proprietary, etc.

Select the Stamp Tool, as shown in Figure 2.112. Select a stamp from the menu, and click to place it on the drawing, as shown in Figure 2.113. You can drag the corners of the stamp to resize it.

FIGURE 2.112

Note that a stamp becomes permanent when the eDrawing is saved, so the recipients cannot delete it.

There are several options for saving and sending eDrawings files. Files can be e-mailed directly from the eDrawings Publisher.

Select the Send Tool, as shown in Figure 2.114.

FIGURE 2.113

FIGURE 2.114

Several options are available, as shown in **Figure 2.115**. To send the drawing to someone without the eDrawings Viewer, it is necessary to save an executable file, which includes the Viewer. Since the executable file contains the eDrawing Viewer, the file is much larger than the file saved with the .edrw format. In this example, the .exe file is 4 MB, as compared with the 76 kB file size of the .edrw file. Since many companies have firewalls that prevent .exe files from being received (because of the possibility of viruses), it may be necessary to send the file as a zipped file or as an html page. Both of these options result in the eDrawings Viewer being sent along with the eDrawing.

FIGURE 2.115

Select Cancel to close the Send box. Select File: Save from the main menu, save the eDrawing as an .edrw file, and close the eDrawings Publisher.

PROBLEMS

P2.1   Create a SolidWorks drawing of the beam section described in Problem P1.2. Include Front and Right Views in your drawing.

P2.2   Create a SolidWorks drawing of the pulley created in Chapter 1. Include Front and Right Views in your drawing, as well as a section view. Show a detail view of the keyway region.

P2.3   Create an eDrawing of the pulley from Problem P2.2.

P2.4 Create a SolidWorks drawing of the plastic pipe tee described in Problem P1.5. Include Front and Right Views in your drawing, as well as a section view.

FIGURE P2.4

P2.5 Create an executable eDrawing of the pipe tee from Problem P2.4. (This problem requires the eDrawing add-in.)

P2.6 Create a SolidWorks drawing of the brake rotor described in Problem P1.6. Include Front and Right Views in your drawing.

FIGURE P2.6

P2.7 Create an eDrawing of the brake rotor from Problem P2.6.

P2.8   Create a SolidWorks drawing of the connecting rod described in Problem P1.10.

P2.9   Create an eDrawing of the connecting rod from Problem P2.8.

P2.10   Problem P1.11 describes a required modification to the connecting rod of Problem P1.10. Using the eDrawing Markup Tool, add a markup to the eDrawing of Problem P2.9 indicating to the designer that the weight reduction modification is necessary and recommending the area for weight removal.

CHAPTER 3

Additional Part Modeling Techniques

Introduction

The flange part modeled in Chapter 1 required the use of a number of basic solid modeling construction techniques. In this chapter, some additional construction and dimensioning techniques will be explored. The part that will be modeled in this chapter is a wall-mounted bracket, as shown in **Figure 3.1**.

FIGURE 3.1

This bracket has some features that will allow us to explore some new modeling techniques:

- It has one plane of symmetry, allowing us to exploit the mirroring capabilities of the program.

- It has a linear hole pattern, which can be readily constructed using the SolidWorks program.

- It has three reinforcing ribs, which can be created with the extruded boss command used earlier or with a rib tool.

In this chapter, a tutorial for modeling this bracket will be presented. Before beginning this tutorial, we will learn how to customize the SolidWorks environment.

Chapter Objectives

In this chapter, you will:

- learn how to customize the SolidWorks interface,

- learn how to use toolbars and keyboard shortcuts,

- use symmetry in part modeling,

- learn to use the Rib Tool,

- create a linear hole pattern,

- create an eDrawing of a part, and

- learn how to create an animation of a part.

3.1 Customizing and Using the SolidWorks Environment

In Chapters 1 and 2 we learned how to modify system settings in SolidWorks and how to modify document-specific settings and save them in templates. In this section, we will introduce some of the ways in which the SolidWorks User Interface can be customized to your preferences.

Open a new part file.

First, let's consider the CommandManager. The CommandManager was added to the SolidWorks program beginning with the 2004 release. Prior to that release, commands were made from toolbars, and many experienced users prefer to use toolbars.

Click the arrow beside the Options Tool and select Customize, as shown in Figure 3.2. Clear the check box labeled "Enable CommandManager," as shown in Figure 3.3, and click OK.

FIGURE 3.2   **FIGURE 3.3**

Note that the Features and Sketch Toolbars are automatically opened, since those tabs were active on the CommandManager. Other toolbars can be added.

Select Customize again, and scroll down the list of toolbars to Standard Views. Click the check box, as shown in Figure 3.4, and click OK.

FIGURE 3.4

Note that the list of toolbars can also be accessed by right-clicking on any toolbar or the Command-Manager, and the Customize menu can be accessed from the bottom of the list of toolbars.

FIGURE 3.5   **FIGURE 3.6**

The toolbars can be moved to other locations by clicking and dragging the move handle at the top or left of the toolbar if it is along the edge of the screen (**Figure 3.5**) or by clicking and dragging the toolbar's title bar if it is floating within the graphics area (**Figure 3.6**).

Experiment with moving the toolbars. Place them in the locations shown in Figure 3.7.

The main advantage of using toolbars is that all of the tools can be accessed with a single mouse click, while with the CommandManager you often have to switch to the desired group of tools before selecting the tool. The disadvantage, at least for new users, is that the name of the tool, which is displayed for many of the tools in the CommandManager, is not visible. However, if the "Show tooltips" box is checked in the Customize menu, then a description of each tool still appears when you hold the cursor over that tool, as shown in **Figure 3.8**. (If the "Show large tooltips" box is checked, then the full description is shown, as in **Figure 3.8**. If that box is unchecked, then only the name of the tool is displayed.)

The toolbars themselves can be edited. To demonstrate this, consider the Centerline Tool. To access this tool from either the Command-Manager or the default configuration of the Features Toolbar, it is necessary to click on the arrow beside the Line Tool and select the Centerline Tool from the menu that appears. If you use this tool often, then you may desire to have one-click access to it.

Right-click on one of the toolbars and select Customize from the menu. Select the Commands tab, as shown in Figure 3.9.

Commands are grouped by categories. The first category, "Flyout Toolbars," contains the tools that have pull-down menus associated with them. For example, the Line Tool in the Flyout Toolbars group has a menu for selecting either the Line or Centerline commands, while the Line Tool in the Sketch group executes only the Line command.

FIGURE 3.7

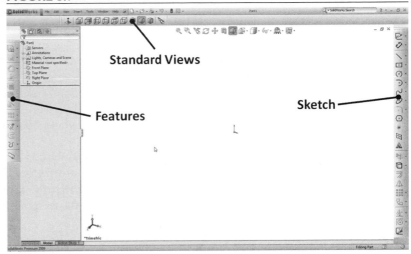

Standard Views

Features

Sketch

FIGURE 3.8

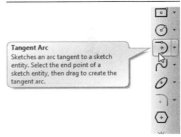

Tangent Arc
Sketches an arc tangent to a sketch entity. Select the end point of a sketch entity, then drag to create the tangent arc.

FIGURE 3.9

FIGURE 3.10

Click on the Sketch Group and locate the Centerline Tool, as shown in Figure 3.10. Click and drag the tool to a location on the Sketch Toolbar immediately below the Text Tool, as shown in Figure 3.11. The plus sign indicates that the tool will be added. Release to place the tool, as shown in Figure 3.12.

Tools that are rarely used can be removed from a toolbar. For example, if you rarely use 3-D sketches, you may want to remove the Plane Tool from the Sketch Toolbar.

Locate the Plane Tool, as shown in Figure 3.13. Click and drag the Plane Tool into the graphics area, as shown in Figure 3.14. The X indicates that the tool will be deleted. Release to delete the tool. Click OK to close the Customize box.

FIGURE 3.11   **FIGURE 3.12**   **FIGURE 3.13**   **FIGURE 3.14**

FIGURE 3.15

Note that if you click Cancel instead of OK, the changes that you made will still take effect. Therefore, if you delete a tool by mistake, you will have to add it back using the procedure described previously.

Remember that if you delete a command from a toolbar or the CommandManager, you can still access that command from the main menu. For example, the Plane Tool can be accessed by selecting Tools: Sketch Entities: Plane, as shown in **Figure 3.15**.

The CommandManager can also be customized.

Click the arrow beside the Options Tool and select Customize. Click the check box labeled "Enable Command-Manager," as shown in Figure 3.16.

Note that the Features and Sketch Toolbars are closed since those commands are now included in the CommandManager, while the Standard Views Toolbar remains open.

FIGURE 3.16

Click the check box beside Standard Views to turn off display of that toolbar. Select the Commands tab. Click on the Sketch Group. Click and drag the Centerline Tool to the Sketch group of the CommandManager, immediately below the toolbar between the Plane and Text Tools, as shown in Figure 3.17.

The Centerline Tool now appears in the CommandManager, as shown in **Figure 3.18**. Note that some of the tools in the CommandManager are displayed with text labels, while some are not. Display of the text can be set for each tool.

Right-click on the Centerline Tool. Click Show Text. Right-click again and click Text Below.

The Centerline Tool now appears as shown in **Figure 3.19**.

Click and drag the Centerline Tool into the graphics area, as shown in Figure 3.20, to remove it from the CommandManager.

Any of the toolbars can be displayed on the CommandManager.

Click on the New Tab icon at the end of the CommandManager tabs, as shown in Figure 3.21. From the list of toolbars, select Standard Views.

FIGURE 3.17

FIGURE 3.18

FIGURE 3.19

FIGURE 3.20

FIGURE 3.21

FIGURE 3.22

The Standard Views tools are now displayed as a group on the CommandManager, as shown in **Figure 3.22**. You can also add a blank tab to the CommandManager, populate it with any combination of tools, and name it. Most users will consider adding a custom tab to be unnecessary, as the Sketch and Features Toolbars contain the majority of the tools used on a regular basis, and both have room for the addition of a few other tools.

FIGURE 3.23

Right-click on the Standard Views tab and select Delete, as shown in Figure 3.23. Click OK to end the customization process.

Another way of selecting tools is through the Shortcut Bar. This context-sensitive menu is accessed by pressing the S key on the keyboard. The following steps demonstrate its use.

Click on the Front Plane to select it. Click the Sketch icon on the pop-up toolbar, as shown in Figure 3.24. In the graphics area, press the S key. From the menu that appears, select the Circle Tool, as shown in Figure 3.25. Drag out a circle. Press the S key, and select the Smart Dimension Tool, as shown in Figure 3.26. Add a diameter dimension to the circle. Press the S key, and select the Exit Sketch Tool, as shown in Figure 3.27. Press the S key, and select the Extruded Boss/Base key, as shown in Figure 3.28. Click the check mark in the PropertyManager to complete the extrusion.

FIGURE 3.24

FIGURE 3.25

FIGURE 3.26

FIGURE 3.27

FIGURE 3.28

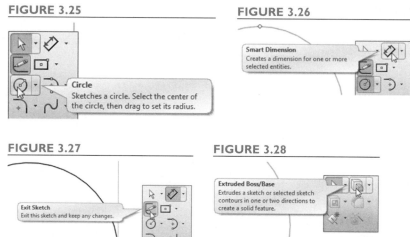

Note that the tools displayed were Sketch tools when a sketch was active, and Features tools otherwise. The tools in the Shortcut Bar can be customized by right-clicking in the bar and dragging tools into and out of the bar in a similar manner to that of the toolbars and CommandManager. For example, the Centerline Tool can be added, as shown in Figure 3.29. Note that it is necessary to have a sketch open to edit this set of tools; otherwise, the Features tools will be displayed and can be edited.

FIGURE 3.29

Accessing the Shortcut Bar by pressing the S key is an example of a keyboard shortcut. There are many other keyboard shortcuts available; a complete list can be found by selecting the Customize Tool and clicking on the Keyboard tab. Some of the more useful shortcuts are listed here.

	Key(s)	Command
Viewing Tools		
	F	Zoom to Fit: allows the entire model to be displayed
	G	Magnifying Glass: creates a magnifier, allowing small details to be seen without changing the overall zoom level
	Z	Zoom Out
	Shift+Z	Zoom In
	C	Expand/Collapse Tree: causes the fly-out FeatureManager to display as collapsed or fully expanded
	Space Bar	Orientation: Displays a menu of Standard View options
Selection Tools		
	X	Filter Faces: allows only model faces to be selected
	E	Filter Edges: allows only edges to be selected
	F6	Toggle Filters: turns on/off selected filters
Other		
	S	Shortcut Bar

A note about the selection tools, which are used most frequently when mating parts within an assembly: when a selection filter is active, then a filter icon is displayed beside the cursor, as shown in **Figure 3.30**. When this icon is displayed, then only the type of entity specified by the filter can be selected. Occasionally, a new user will press a key by accident (most likely the X key), which causes a filter to become active, preventing any other selections. If this happens, then pressing the F6 key clears the filter.

FIGURE 3.30

In this section, we have seen that the SolidWorks User Interface allows for multiple ways to select commands and for a great deal of customization to fit a user's preferences. In this book, we will mostly use the CommandManager and the Heads-Up Toolbar to select commands, with a minimal amount of customization (such as adding the Rotate View, Pan, and Trimetric View Tools to the Heads-Up Toolbar), and we recommend that new users limit customization. Experienced users will no doubt find many individual preferences for setting up and using the interface.

Close the part. It is not necessary to save the part file.

3.2   Part Modeling Tutorial: Bracket

Open a new part file. Be sure that the dimensions are set to inches, with three decimal places displayed.

FIGURE 3.31

Click on the Front Plane in the FeatureManager to select it, as shown in Figure 3.31.

From the Sketch group of the CommandManager, select the Center Rectangle Tool, as shown in Figure 3.32.

In the drawing area, snap to the origin, as shown in Figure 3.33. This will locate the center of the rectangle at the origin. Left-click and drag out the rectangle, as shown in Figure 3.34.

FIGURE 3.33

Don't worry about the locations of the corner points; we will add dimensions that will size the rectangle.

FIGURE 3.32

FIGURE 3.34

Click on the Smart Dimension Tool, and dimension a horizontal side of the rectangle to 6 inches and a vertical side to 4 inches, as shown in Figure 3.35.

The rectangle is now both dimensioned and located, so the sketch is fully defined.

Click on the Features tab and then the Extruded Boss/Base Tool. In the PropertyManager, enter 0.25 inches as the thickness, and click on the check mark, as shown in Figure 3.36.

FIGURE 3.35

FIGURE 3.36

A solid object has been created, as shown in **Figure 3.37**.

FIGURE 3.37

The horizontal boss of the bracket will now be constructed. The feature will be sketched in a plane 1.25 inches below the top surface of the base part. To accomplish this, we will first need to create the reference plane to be used for sketching.

Select the top surface of the base part (see Figure 3.38).

Be sure that the face is selected (a square symbol will appear next to the cursor prior to selection).

FIGURE 3.38

From the Features group of the CommandManager, select Reference Geometry: Plane, as shown in Figure 3.39. In the PropertyManager, click the check box to reverse the direction, and set the distance to 1.25 inches (see Figure 3.40). Click the check mark to create the plane, which will be named Plane1.

FIGURE 3.39

FIGURE 3.40

Note that the boundaries of the plane are similar in size and shape to those of the top surface. Of course, a plane extends infinitely, so these boundaries can be adjusted for better visibility.

Click and drag the move handles on Plane1 to change the plane's visible boundaries as desired, as shown in Figure 3.41.

The boss feature will now be sketched in this new plane. The symmetry of the boss about the Right Plane will be exploited in the creation of the sketch.

FIGURE 3.41

Switch to the Top View. Use the Pan Tool to place the part toward the top of the screen, as shown in Figure 3.42. Press the Esc key to turn off the Pan Tool.

FIGURE 3.42

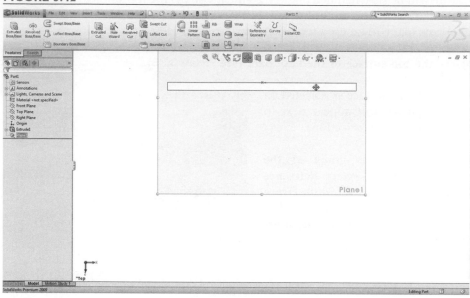

We will use mirroring to assist in construction of the symmetric sketch. This will require that we draw a centerline first.

If Plane1 is not selected (highlighted), then click on Plane1 in the FeatureManager to select it. Click the Sketch tab of the CommandManager to display the Sketch tools. Select Centerline from the pull-down menu associated with the Line Tool. Drag a centerline from the origin downward, as shown in Figure 3.43. As you drag the line downward, a vertical relation symbol by the cursor shows that the line will be vertical.

FIGURE 3.43

Select the Line Tool from the Sketch group of the CommandManager. Move the cursor near the lower edge of the solid rectangle, as shown in Figure 3.44.

A coincident relation symbol will appear next to the cursor, indicating that you are snapping to the edge of the solid.

Click and hold, while dragging a line downward and at an angle (Figure 3.45).

Select the Tangent Arc Tool, as shown in Figure 3.46. Move the cursor to the end of the line, snapping the start of the arc to this endpoint. Hold down the left mouse button and drag out an arc, with the endpoint of the arc snapped to the centerline, as shown in Figure 3.47. Press the Esc key to turn off the Tangent Arc Tool.

FIGURE 3.44

FIGURE 3.45

FIGURE 3.46

FIGURE 3.47

The arc that you drew will remain tangent to the first line, no matter how the dimensions of the sketch are changed.

We will now exploit the symmetry of the bracket, and construct a mirror image of our sketch about the centerline.

With the Ctrl key depressed (to allow for multiple selections), click to select the centerline, the solid line, and the tangent arc (Figure 3.48). Select Mirror Entities from the CommandManager, as shown in Figure 3.49. A symmetric sketch will be created about the selected centerline, as shown in **Figure 3.50.**

FIGURE 3.48

FIGURE 3.49

While this sketch does exhibit the required symmetry, it does not capture our design intent. We need to add an additional relation between the two tangent arcs.

Select the two arcs, and click Concentric in the PropertyManager, as shown in Figure 3.51. Click the check mark to add the relation.

FIGURE 3.50

FIGURE 3.51

The two arcs are now tangent to each other, as shown in **Figure 3.52.** Note that we could have specified a tangent relation instead of the concentric relation, but there are two possible solutions to the tangent relation. To ensure that the geometry is the way we want it, the concentric relation provides only one solution.

FIGURE 3.52

Select the Line Tool, and finish the sketch with a horizontal line connecting the two free ends of the sketch, as shown in Figure 3.53. Select the Smart Dimension Tool, and dimension the sketch as shown in Figure 3.54. Add the 0.75 inch radius first.

FIGURE 3.53

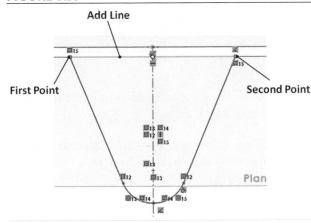

Add Line

First Point

Second Point

Plan

FIGURE 3.54

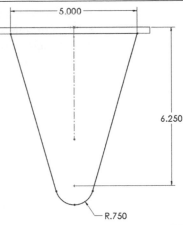

5.000

6.250

R.750

FIGURE 3.55

Note that the 6.25-inch dimension is to the back of the bracket.

Click on the Features tab of the CommandManager. Select the Extruded Boss/Base Tool. Click the Reverse Direction icon in the PropertyManager, as shown in Figure 3.55, and extrude the part downward 0.25 inches. Change to the Trimetric View to preview the extrusion, and click the check mark to complete the operation.

The bracket should appear as shown in **Figure 3.56**.

Now we will add a raised cylindrical feature for reinforcement near what will become a mounting hole.

Select the top surface of the horizontal extruded boss, as shown in Figure 3.57. Select the Normal To View.

FIGURE 3.56

FIGURE 3.57

We want to locate the raised area concentric with the rounded tip of the boss. In order to establish this relation, we will first need to "wake up" the arc that makes up this rounded tip, allowing the center point to be used as a snap point in our sketch.

FIGURE 3.58

Click the Sketch tab of the CommandManager. Select the Circle Tool. Without clicking, drag the cursor until the tip of the pencil icon touches the arc, as shown in Figure 3.58 **(which is shown in wireframe mode for clarity).**

The center point of the arc now appears, and it can be used as a snap point for the center of the circle.

FIGURE 3.59

Drag out a circle, starting at the center mark. Select the Smart Dimension Tool, and dimension the circle to be 1 inch in diameter (see Figure 3.59**).**

Select the Extruded Boss/Base Tool from the Features group of the CommandManager. Extrude the sketch upward for a distance of 0.050 inches, as shown in Figure 3.60.

A reinforcing rib will now be added to the part. This will be created using a new type of extrusion, called a Mid-Plane Extrusion.

FIGURE 3.60

Select the Right Plane from the FeatureManager, and select the Right View. Select the Line Tool from the Sketch group of the CommandManager, and sketch the line shown in Figure 3.61. **The endpoints of your line should snap to the edges of the solids, but not to any specific point on the edges. Recall that a coincident relation symbol next to the cursor indicates that you are snapping to the edge.**

Add one horizontal and one vertical line coincident with the edges of the solid (snap to the endpoints of the first line and the intersection point of the edges), completing the profile of the rib, as shown in Figure 3.62.

FIGURE 3.61

FIGURE 3.62

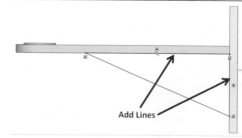

Select the Smart Dimension Tool, and dimension the sketch as shown in Figure 3.63.

FIGURE 3.63

Select the Extruded Boss/Base Tool from the Features group of the CommandManager. In the Property-Manager, select Mid Plane as the type from the pull-down menu (see Figure 3.64), and enter a thickness of 0.1875 inches (see Figure 3.65). Click the check mark to complete the extrusion.

Unlike the one-sided blind and "up to next" extrusions used previously, the mid-plane type extrudes material in *both* directions, from the sketch plane outward. For a Mid-Plane Extrusion, the thickness entered is the *total* thickness of material after the extrusion. A front view of the part shows that the rib is symmetric about the Right Plane (see Figure 3.66).

FIGURE 3.64

FIGURE 3.65

FIGURE 3.66

The next rib will be added in a slightly different manner, using the Rib Tool. But first, we must create a new plane for the rib. So far, we have used the CommandManager to create a plane. Here, we will use a drag-and-drop technique that is similar to that used in other Windows applications.

Select the Trimetric View. Click on the Right Plane in the FeatureManager to select it. Hold down the Ctrl key, and move the cursor over one of the lines defining the plane until the move arrows appear, as shown in Figure 3.67. Click and drag the mouse to the right. When the mouse button is released, a new plane is created, as shown in Figure 3.68. In the PropertyManager, set the offset distance to 1 inch, as

FIGURE 3.67

FIGURE 3.68

FIGURE 3.69

shown in Figure 3.69. **Click the check mark to complete the definition of the new plane, which will be labeled Plane2.**

Select the Line Tool from the Sketch group of the CommandManager. Draw a diagonal line as shown in Figure 3.70, snapping the endpoints to the edges of the solids (but not to any specific points). Add the two dimensions shown.

For the first rib, we added two lines to close the sketch and used an Extruded Boss command to complete the rib. A rib can be created from an open sketch using the Rib Tool.

Select the Rib Tool from the Features group of the CommandManager, as shown in Figure 3.71. In the FeatureManager, set the thickness to Both Sides and 0.1875 inches, as shown in Figure 3.72.

FIGURE 3.70

FIGURE 3.71

FIGURE 3.73

FIGURE 3.72

The thickness can be set to be offset from the plane in either direction, or so that the rib is symmetric about the plane (similar to the Mid-Plane Extrusion used for the first rib). By default, the middle of the three buttons, designating a rib symmetric about the plane, is selected. For the extrusion direction, the rib can be defined as parallel to the plane of the sketch (default) or normal to the plane. Neither of these settings needs to be changed for our rib.

FIGURE 3.74

Check the "Flip material side" box, as shown in Figure 3.72, so that the arrow defining the rib direction points toward the part, as shown in Figure 3.73. Click the check mark to complete the rib, which is shown in Figure 3.74.

Earlier, we used the Mirror Entities Tool to create symmetric elements within a sketch. A similar tool in the Features group of the CommandManager, called the Mirror Tool, is used to create symmetric features in a part. We will use this tool to create the final rib.

Click on the Features Tab of the CommandManager. Select the Mirror Tool, as shown in Figure 3.75. In the PropertyManager, the Mirror Face/Plane box is active (highlighted), as shown in Figure 3.76.

FIGURE 3.75

FIGURE 3.76

Click on the Right Plane in the FeatureManager, which has "flown out" to the right of the PropertyManager. It may be necessary to click on the plus sign next to the part name to expand the FeatureManager before selection. The rib just created should be listed as the feature to mirror. If it is not, click that box to make it active and select the rib from the FeatureManager.

A preview of the final rib is displayed, as shown in Figure 3.77.

Click the check mark to complete the rib.

The three ribs are shown in Figure 3.78.

FIGURE 3.77

FIGURE 3.78

The holes will now be added to the part.

Select the top surface of the raised circle, as shown in Figure 3.79. Select the Circle Tool from the Sketch group of the Command-Manager. Select the Top View.

FIGURE 3.79

As before, it is necessary to "wake up" a center mark at which the hole will be centered.

Move the cursor to the edge of the rounded end of the part, or the circular edge of the reinforced area, without clicking a mouse button. This will cause the center mark to be displayed, as shown in Figure 3.80.

Click and drag out a circle from the center mark. Select the Smart Dimension Tool, and dimension the diameter of the circle as 0.50 inches, as shown in Figure 3.81.

FIGURE 3.80

FIGURE 3.81

Ø.500

Select the Extruded Cut Tool from the Features group of the CommandManager, and select Through All as the type. Click the check mark to complete the hole, which is shown in Figure 3.82.

FIGURE 3.82

The four-hole bolt hole pattern used to mount the bracket to the wall will now be added. This will be done by creating a single hole, and then defining a *pattern* based on this hole. In Chapter 1, a *circular pattern* was used to create a bolt hole pattern in the flange. In this exercise, a new type of pattern known as a *linear pattern* will be introduced.

The first step will be the creation of a single bolt hole.

Select the large vertical face of the base part as shown in Figure 3.83 and select the Normal To View. Select the Circle Tool from the Sketch group of the CommandManager, and sketch a circle near the upper-left corner of the face. Select the Smart Dimension Tool and dimension the hole diameter as 0.375 inches, as shown in Figure 3.84.

FIGURE 3.83

FIGURE 3.84

FIGURE 3.85

Add linear dimensions between the circle and the edges, as shown in Figure 3.85.

Select the Extruded Cut Tool from the Features group of the Command-Manager, and select Through All as the type. Click the check mark to complete the hole, which is shown in Figure 3.86.

With this single hole defined, a linear hole pattern can now be created.

FIGURE 3.86

With the new hole selected, select the Linear Pattern tool from the Features group of the CommandManager, as shown in Figure 3.87.

FIGURE 3.87

DESIGN INTENT | Symmetry in Modeling

In this chapter, we used mirroring techniques for sketch entities and features to define symmetric elements in a part. The four mounting holes in the bracket, which were created with a linear pattern, could have been defined with mirror commands. The advantage of doing so would be that if we change the dimensions of the vertical portion of the bracket (6 inches by 4 inches), the hole positions would be updated to the correct positions. The same associativity between the hole locations and the size of the bracket can be added if a linear pattern is used, but

will require the addition of equations, which will be explained in Chapter 4.

Using symmetric elements in a part is good modeling practice. If a part contains planes of symmetry, plan your model to take advantage of those planes. In the bracket model, we centered the initial sketch about the origin. This placed the part so that the Right Plane coincided with the symmetry plane of the entire part, and the Top Plane coincided with an additional symmetry plane of the vertical portion of the part.

The feature Extrude6 will be shown in the "Features to Pattern" box, indicating that this is the base feature for the linear pattern. The "Direction 1" box in the PropertyManager will be highlighted; this allows the first linear direction of the pattern to be established.

FIGURE 3.88

Click on the top horizontal edge of the rectangular base to establish the first direction, as shown in Figure 3.88. In the PropertyManager, set the Direction 1 distance to 5 inches and the number of instances to 2, as shown in Figure 3.89. The Direction 2 box in the PropertyManager will be highlighted.

A preview of a second patterned hole will appear, as shown in Figure 3.90.

FIGURE 3.89

Click on the right vertical edge of the rectangular base to establish the second direction of the pattern, as shown in Figure 3.91.

FIGURE 3.90

FIGURE 3.91

FIGURE 3.92

In the PropertyManager, set the **Direction 2 distance** to be 3 inches and the number of instances to be 2, as shown in Figure 3.92.

If the preview of the patterned holes shows that the pattern is created in the wrong direction, click on the arrow at the end of the edge defining the pattern direction, as shown in Figure 3.93.

When the preview of the pattern appears correct, as shown in Figure 3.94, click the check mark to complete the pattern.

The completed part is shown in Figure 3.95.

From the FeatureManager, right-click **Material** and select **Edit Material**, as shown in Figure 3.96. Select **ABS** from the **Plastics** group, as shown in Figure 3.97. Click **Apply**, and then **Close**.

Save the part file.

FIGURE 3.93

FIGURE 3.94

FIGURE 3.95

FIGURE 3.96

FIGURE 3.97

3.3   Sharing and Displaying the Solid Model

In Chapter 2, we created an eDrawing of a 2-D drawing. You can also create an eDrawing of a part. An eDrawing can be viewed by anyone who downloads the free eDrawing viewer. The small file sizes of eDrawings make them easy to share via e-mail.

FIGURE 3.98

Select Tools: Add-Ins from the main menu. Click the check box to the left of eDrawings 2009, as shown in Figure 3.98.

Note that if you want eDrawings to be active every time you open SolidWorks, then you can check the box on the right side.

Click OK. From the main menu, select File: Publish eDrawings 2009 File, as shown in Figure 3.99.

FIGURE 3.99

The eDrawings program will open, and the bracket model will be displayed, as shown in **Figure 3.100**. Although the eDrawings model does not contain the history of the steps that were used to create it, and cannot be edited, it does contain the geometric data sufficient to view the model from various angles and view a cross-section, and it also contains the mass properties data.

FIGURE 3.100

FIGURE 3.101

Click the Next button several times, as shown in Figure 3.101.

The Model View will shift from the Default View from the SolidWorks file (trimetric) to isometric, and then to each of the principal views. The Play button will automatically rotate the part through the various views until the Stop button is pressed. The Home button returns the view to the default. You can zoom in and out with tools similar to those of SolidWorks.

Select the Home button, and then the Section Tool, as shown in Figure 3.102. At the left of the screen, choose the YZ Plane as the section plane, as shown in Figure 3.103.

FIGURE 3.102

FIGURE 3.103

Note that you can click and drag the plane through the part, as shown in Figure 3.104.

FIGURE 3.104

Click the Section Tool again to turn it off, and select the Mass Properties Tool, as shown in Figure 3.105.

The mass, volume, and surface area are displayed, as shown in Figure 3.106. Note that you can select more decimal places to be displayed if desired.

FIGURE 3.105

Select Save, and save the eDrawing file to the desired location.

FIGURE 3.106

Note that the file size is about the same as that of a small word-processing file. This makes the file ideal for sharing via e-mail. The Mail Tool in eDrawings allows the file to be sent in several other formats that include the eDrawings viewer. These formats are discussed in more detail in Chapter 2.

Close eDrawings, and return to SolidWorks.

Often, you will want to display your solid model in a text document or in a presentation. The easiest way to do this is with screen captures of the model.

From the main menu, select View: Screen Capture: Image Capture, as shown in Figure 3.107. Open a Word document, and select Paste.

FIGURE 3.107

The image is copied to the Windows clipboard and then pasted into the Word (or PowerPoint) document. While the Prnt Scrn key performs a similar function, an advantage of SolidWorks Image Capture is that only the graphics area is captured. With the Prnt Scrn key, the entire screen is captured and must be cropped if only the model is desired in the image.

The menu also includes a Record Video option, which records the screen as the model is rotated on the screen. However, we can create animations with the MotionManager, which gives us better control over the manipulation of the model.

FIGURE 3.108

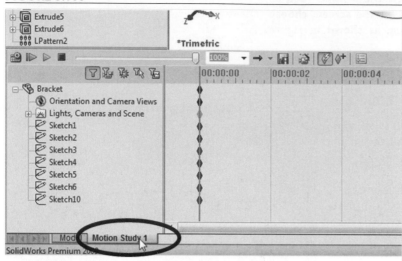

Select the Motion Study 1 tab at the bottom of the screen, as shown in Figure 3.108. (If the tab is not visible, then select View: MotionManager from the main menu.)

A timeline is displayed. We will set the view orientations of the model that we desire along the timeline, using *keys*.

Select the Trimetric View. Zoom and pan the model as desired. Since the recorded video will not show the FeatureManager, placing the model slightly to the left of center in the graphics area will result in the model being centered left-to-right in the video. Right-click the diamond-shaped key beside Orientation and Camera Views, and select Replace Key, as shown in Figure 3.109. Move the cursor to the 2-second mark on the timeline, with Orientation and Camera Views still highlighted. Right-click and select Place Key, as shown in Figure 3.110.

FIGURE 3.109

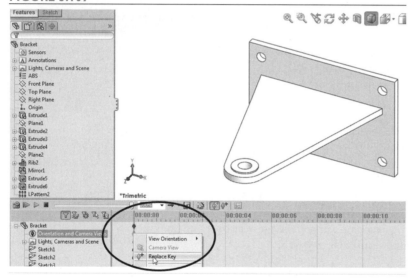

The purpose of placing this second key is to display the model in the selected orientation for two seconds before beginning to rotate it. Since the ribs are best seen from the Bottom View, we will now rotate the model to that view.

FIGURE 3.110

Select the Bottom View. Zoom and pan the model as desired, again placing it to slightly left of center in the graphics window. Move the cursor to the 4-second mark, to the right of Orientation and Camera Views, right-click and select Place Key. Repeat for the 6-second mark.

The timeline will now show a line between the keys at 2 and 4 seconds. As shown in **Figure 3.111**, this indicates that the views are different at these two points, and so the model will be rotated into the new view over this 2-second interval.

FIGURE 3.111

Select the Right View and zoom and pan the model as desired. Place keys at 8 and 10 seconds. Select the Trimetric View, zoom and pan as desired, and place keys at 12 and 14 seconds.

The finished timeline is shown in **Figure 3.112**. Note that individual keys can be deleted or replaced as necessary by right-clicking on that key. To start over, you can right-click a key, choose Select All, and press the Delete key.

FIGURE 3.112

To play the animation, select Play from Start, as shown in Figure 3.113.

We will now save the animation as a video file. First, we will minimize the Motion-Manager so that the aspect ratio of the graphics area is better for the finished video.

Collapse the MotionManager by selecting the double arrows at the right edge of the screen, as shown in Figure 3.114. Select the Save Animation Tool, as shown in Figure 3.115. Select the file name and directory, and accept the default settings.

FIGURE 3.113

FIGURE 3.114

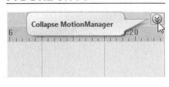

The animation will run on the screen as the file is created. When the animation is completed, the saved video file can be viewed from a standard player, as shown in **Figure 3.116**. Videos can also be embedded into PowerPoint presentations. This allows a model to be displayed and rotated during the presentation without having to open and run SolidWorks on the presentation computer.

FIGURE 3.115

Save Animation
Saves the animation as an AVI or other file type.

FIGURE 3.116

Save and close the SolidWorks file.

PROBLEMS

P3.1 Create a solid model of a structural tee section by creating the sketch shown in
Figure P3.1A on the Front Plane and mirroring it about the centerline. Use relations
as appropriate to fully define the sketch. Extrude it using a Mid-Plane Extrusion to a
total thickness of 36 inches to create the model shown in **Figure P3.1B**.

FIGURE P3.1A   **FIGURE P3.1B**

P3.2 The cross-section of a W14 × 370 Wide-Flange Beam is shown in **Figure P3.2A**.
Create a model of a 36-inch long segment of a W14 × 370 beam by creating the
sketch shown in **Figure P3.2B** and mirroring it about the two centerlines. (Note: This
must be done in two steps. Select the items to be mirrored and one of the
centerlines, and then select the Mirror Entities Tool. Then select all entities to be
mirrored and the other centerline and select the Mirror Entities Tool. Only one
centerline can be selected for each mirror operation.) Extrude the sketch to yield the
beam section shown in **Figure P3.2C**. Set the material type to Plain Carbon Steel and
find the weight of the beam segment.

(Answer: 1101 lb)

FIGURE P3.2A   **FIGURE P3.2B**   **FIGURE P3.2C**

P3.3 Modify the wide-flange beam segment of P3.2 by adding six 1-inch-diameter holes to each end of the beam, as shown in **Figure P3.3A**. The locations of the holes are shown in **Figure P3.3B**. Create a single hole, and use a linear pattern to place the other five holes in one end. Then use a mirror command to place the holes in the other end.

FIGURE P3.3A   **FIGURE P3.3B**

P3.4 Create a part model of the plastic bracket shown in **Figure P3.4A** and detailed in **Figure P3.4B**. Use symmetry in your model so that if you change the width of the part from 2 to 3 inches, the rib and hole placements remain symmetric, as shown in **Figure P3.4C**.

FIGURE P3.4A   **FIGURE P3.4B**

FIGURE P3.4C

P3.5 Create a model of the perforated board shown in **Figure P3.5A** by the following procedure:

    a.   Begin with a 4-inch by 2-inch sketch, centered at the origin (**Figure P3.5B**).

    b.   Extrude the sketch 0.1 inch.

    c.   Create and locate a square thru-hole, as shown in **Figure P3.5C**.

    d.   Use a linear pattern to create an evenly spaced 40 × 20 grid of holes.

FIGURE P3.5A

FIGURE P3.5B

FIGURE P3.5C

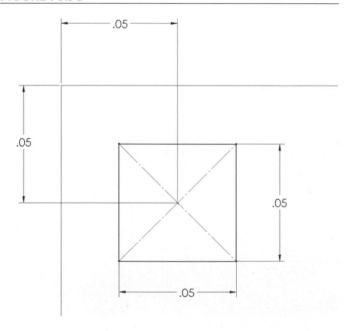

P3.6 Create a solid model of a hacksaw blade (as shown in **Figure P3.6A**, with a close-up view of the teeth shown in **Figure P3.6B**), using a linear patterned cut to create the saw teeth. Follow the procedure outlined below.

    a.   Begin by creating a sawblade "blank," using the dimensions shown in **Figure P3.6C** and extruding the shape to a 0.02-inch depth.

    b.   Sketch a single tooth profile, and extrude a cut through the blank (see **Figure P3.6D**).

    c.   Create a linear pattern to copy the tooth profile the length of the sawblade.

FIGURE P3.6A

FIGURE P3.6B

FIGURE P3.6C

FIGURE P3.6D

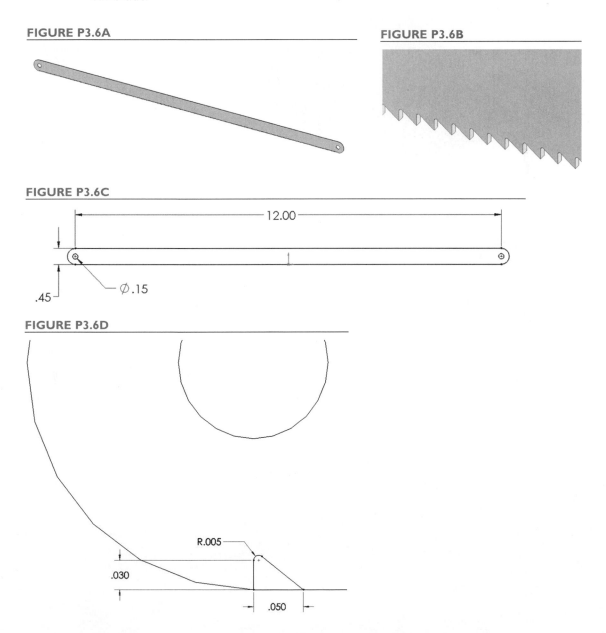

P3.7   Create a solid model of the muffin pan shown in **Figure P3.7A** and detailed in **Figure P3.7B**. The entire pan is 0.050 inches thick. Each of the 12 muffin wells is 3 inches in diameter at the top, is 1.5 inches deep, and has tapered sides, as shown in **Figure P3.7C**. The entire top surface and the insides of the muffin wells are to be coated with a nonstick material. Find the surface area that will be coated. (Hint: The Mass Properties Tool gives you the surface area of the entire part, but to get the surface area for a single surface or a specific group of surfaces, select the surface(s), using the Ctrl key to select multiple items, and select Tools: Measure from the main menu.)

FIGURE P3.7A

FIGURE P3.7B

FIGURE P3.7C

P3.8 Create a solid model of the gear rack shown in **Figure P3.8A** using the following procedure:

    a. Begin creating a "gear blank" by extruding a 0.75×0.75-inch cross section to an overall depth of 11.16 inches.

    b. At the left end of the gear blank, create the sketch shown in **Figure P3.8B** (dimensions in inches), and extrude a cut to begin the gear tooth definition.

    c. Add 0.005-inch fillets at the root of the tooth.

    d. Create a linear pattern of both the extruded cut and the fillets to create the final gear profile. The tooth-to-tooth spacing should be 0.215 inches.

FIGURE P3.8A **FIGURE P3.8B**

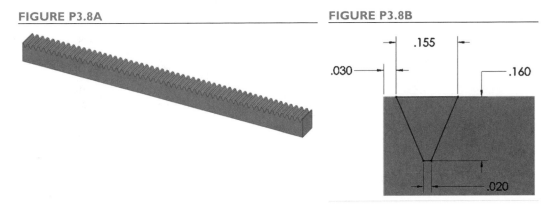

P3.9 Many electronic devices use a heat sink to conduct heat away from a component that may be damaged by high temperatures. A simple heat sink often consists of a conductive metal part with fins. The fins provide a large amount of surface area to allow heat transfer to the air (often with a fan to circulate air over the fins). **Figure P3.9A** shows a copper heat sink; details are shown in **Figure P3.9B**. Model this part, and find the surface area of the part that will be exposed to the air (the bottom surface will be in contact with the electronic component; all other surfaces will be exposed to the air). All dimensions are mm.

FIGURE P3.9A **FIGURE P3.9B**

CHAPTER 4

Use of Parametric Modeling Techniques

Introduction

One of the tremendous advantages of solid modeling software is the ability to represent geometric relationships in a parametric form. In a parametric model, the description of the model contains both a geometric description (shape, interconnectivity, etc.) and specific numerical parameters (dimensions, number of instances, etc.). Consider, for example, a simple parameterized solid model of a cylinder; the definition of the model contains both a geometric description defining the primitive shape (cylinder), and two numeric parameters (radius and length) to fully define the model. A whole "family" of cylindrical parts of different radii and lengths could be modeled with a single parameterized model, with only a numeric table of values used to differentiate the various parts in the family.

The Linear Pattern Tool used to define the hole pattern in Chapter 3 is another good example of a parameterized model; the geometric information used to describe a hole pattern includes:

- the shape of the hole (circular)
- the type of extruded cut ("through all")

The numeric parameters that must also be defined include:

- the vectors (axes) that define the two directions of the pattern
- the two "repeat" dimensions along the two directions
- the number of instances of the holes in each direction

Changes can be readily made to the parameters to update the model, without requiring any changes to the geometric information.

This concept of using numeric parameters to drive a solid model can be exploited in design. One powerful tool that can be employed is the development of mathematical relationships (equations) that relate the values of two or more parameters in a model.

Chapter Objectives

In this chapter, you will:

- learn how to display a section view of a model,

- use the Convert Entities Tool to convert the silhouette of a solid part into a sketch line,

- use the Draft Tool to add draft to an existing solid part,

- learn how to use a shared sketch to create more than one feature,

- develop equations to embed parameterized relationships into a solid model,

- use a design table to create a family of similar parts, and

- create a multiple-configuration part drawing.

117

In our hole pattern, for example, we might want to restrict our model so that the number of instances of holes in one direction is always the same as the number of instances of holes in the other direction. This is accomplished by relating the two parameters together with an equation, for example:

(# of instances along Direction 2) = (# of instances along Direction 1)

Once this parameterized relationship is established, any change made to the *independent parameter* (the number of instances of holes along Direction 1) would automatically update the value of the *dependent parameter* (the number of instances of holes along Direction 2), and the model would be updated accordingly. This allows the engineer to embed design intelligence and "rules-of-thumb" directly into solid models.

In this chapter, an example of a mechanical part that contains embedded equations to drive specific parameters will be presented. In addition, the tutorial will introduce some new modeling operations that have not yet been utilized in the preceding chapters.

4.1   Modeling Tutorial: Molded Flange

In this exercise, we'll create the flange shown in **Figure 4.1**. Since this is to be a molded part, we'll add draft to the appropriate surfaces.

Begin by starting a new part, and selecting the Front Plane from the FeatureManager. Select the Circle Tool from the Sketch group of the CommandManager. Drag out a circle centered at the origin. Select the Smart Dimension Tool and add a 6-inch diameter dimension, as shown in Figure 4.2.

Select the Extruded Boss/Base Tool from the Features group of the CommandManager. Extrude the sketch outward to a thickness of 0.25 inches.

FIGURE 4.1

FIGURE 4.2

FIGURE 4.3

The extruded solid is shown in **Figure 4.3**.

Since the next feature will be drafted, we will define its dimensions in a new plane and extrude its sketch back toward the rest of the part. Therefore, we need to create a new construction plane offset from the Front Plane.

Begin by selecting the Front Plane from the FeatureManager.

The selected plane will be highlighted in green.

While holding down the Ctrl key, click and drag the plane outward. Make sure that the move arrows, as shown in Figure 4.4, appear before clicking. Release the mouse button, and the new plane is previewed, as shown in Figure 4.5.

FIGURE 4.4

FIGURE 4.5

In the PropertyManager, set the offset distance to 1.5 inches, as shown in Figure 4.6, and click the check mark to complete the operation.

FIGURE 4.6

FIGURE 4.7

The new plane will be labeled Plane1 in the FeatureManager. A cylindrical boss will now be extruded with a draft from this new plane back toward the base feature.

Click on the boundary of the new plane (Plane1) and choose the Normal To View from the Context Toolbar, as shown in Figure 4.7, or from the Heads-Up View Toolbar. Select the Circle Tool from the Sketch group of the CommandManager. Drag out a circle from the origin. Select the Smart Dimension Tool and dimension the circle's diameter as 2.2 inches, as shown in Figure 4.8.

FIGURE 4.8

FIGURE 4.9

Reverse
Direction

Select the Extruded Boss/Base Tool from the Features group of the CommandManager, and set the direction such that the boss is extruded toward the base. Change the type of extrusion to Up To Next. (Note: If Up To Next is not available, then the extrusion direction is incorrect.) Turn the draft on. Set the draft angle to 3 degrees and check the "Draft outward" box, as shown in Figure 4.9. Click on the check mark to complete the extrusion. From the Hide/Show Items menu of the Heads-Up View toolbar, select View Planes, as shown in Figure 4.10. This will toggle off the display of planes.

FIGURE 4.10

The result of the drafted extrusion is shown in Figure 4.11.

Select the face shown in Figure 4.12, and select the Normal To View. Select the Circle Tool from the Sketch group of the CommandManager. Drag out a circle centered at the origin. Select the Smart Dimension Tool. Dimension the circle's diameter at 1.8 inches, as shown in Figure 4.13.

FIGURE 4.11

FIGURE 4.12

FIGURE 4.13

Select the Extruded Cut Tool from the Features group of the CommandManager. Select Through All as the type, turn the draft on, and set the draft angle to 2 degrees. Make sure that the "Draft outward" box is checked, as shown in Figure 4.14.

FIGURE 4.14

Click the check mark to apply the cut, which is shown in Figure 4.15.

FIGURE 4.15

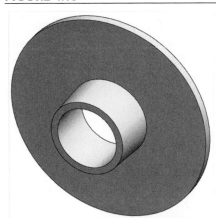

To see the effect of the draft on the hole, viewing the model with a section view is helpful.

Select the Section View Tool from the Heads-Up View Toolbar, as shown in Figure 4.16. In the PropertyManager, select the middle button, representing the Top Plane, as shown in Figure 4.17.

Reverse the direction of the section by clicking on the icon next to the plane selection box.

FIGURE 4.16

FIGURE 4.17

Click the check mark to apply the settings for the section view. Select the Bottom View from the Heads-Up View Toolbar, as shown in Figure 4.18.

The cross-section of the part, as shown in **Figure 4.19**, clearly shows the drafted center hole. The section view will be displayed until turned off with the Section View Tool.

FIGURE 4.18

FIGURE 4.19

Click the Section View Tool to return to the display of the entire model.

Before adding the ribs and holes, we will fillet several of the sharp edges of the part.

FIGURE 4.20

Start by selecting the Fillet Tool from the Features group of the CommandManager, as shown in Figure 4.20.

Since two of the edges are to have the same radius, we can add these fillets at the same time.

Set the radius to 0.125 inches, and select the two edges shown in Figure 4.21.

Be sure to select the edge and not the face. If a face is selected, then all of the edges of that face will be filleted.

Click the check mark to apply the fillets. Using a similar procedure, fillet the edge shown in Figure 4.22 with a fillet radius of 0.25 inches.

The filleted part is shown in **Figure 4.23**.

FIGURE 4.21

FIGURE 4.22

FIGURE 4.23

FIGURE 4.24

A number of stiffening ribs will now be added to the part. This will be done by creating a single rib and duplicating it a number of times using a circular pattern. The first rib in the pattern will now be created.

Select the Top Plane from the FeatureManager. Select the Sketch Tool from the context toolbar, as shown in Figure 4.24, or from the Sketch group of the CommandManager. Switch to the Bottom View.

Note that most of the time when we begin a sketch, we simply choose a sketch tool. However, for the step we are about to perform, it is necessary to open the sketch before choosing the desired tool (Convert Entities).

FIGURE 4.25

Hidden Lines Visible
Displays all edges of the model. Edges
that are hidden from the current view
are displayed in a different color or
font.

Choose the Wireframe display mode with hidden lines visible, as shown in Figure 4.25. Zoom in on the right side of the part.

Since the rib will blend into the fillet radii, we need to select the intersections of the radii with the plane. While the edge of the solid part appears to be an arc when viewed from this perspective, there is no physical sketch entity associated with it; however, we can construct an arc coincident with this projected solid edge by using the Convert Entities Tool.

FIGURE 4.26

Move the cursor to the intersection of the top fillet and the plane. When you have positioned the cursor at the appropriate location, the silhouette symbol will appear (Figure 4.26). Click to select this silhouette.

While holding down the Ctrl key, locate and select the silhouette of the edge of the other fillet, as shown in Figure 4.27.

With the silhouettes selected, click the Convert Entities Tool, as shown in Figure 4.28.

This will add to your sketch arcs that are coincident with the silhouettes of the fillets, as shown in Figure 4.29.

Select the Line Tool. Move the cursor over the top arc so that the coincident relation icon appears, as shown in Figure 4.30, indicating that the endpoint of the line will snap to the arc. Do not snap to the midpoint of the arc.

FIGURE 4.27

FIGURE 4.28

Convert Entities
Converts selected model edges or
sketch entities into sketch segments.

FIGURE 4.29

FIGURE 4.30

Coincident

FIGURE 4.31

FIGURE 4.32

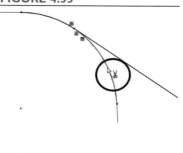

Click and drag a line to the other arc. Place the endpoint so that the coincident and tangent relation icons appear, as shown in Figure 4.31. Press Esc to turn off the Line Tool.

Zoom in on the fillet at the top of the part, as shown in Figure 4.32. A tangent relation should have been added automatically to the arc and the line. If the relation did not add automatically, then add it manually by selecting both entities (use the Ctrl key to select more than one entity) and adding a tangent relation in the PropertyManager. Note that there are also relation icons for On Edge (the arc is on the silhouette edge of the fillet) and coincident (the end of the line lies on the arc).

We will now trim away the portions of the arcs that are not needed in the sketch.

Zoom in on the intersection of one of the arcs and the line. Select the Trim Entities Tool, as shown in Figure 4.33. Choose "Trim to closest" as the trim option, as shown in Figure 4.34. Click to trim the portion of the arc to be removed, as shown in Figure 4.35.

FIGURE 4.33

FIGURE 4.34

FIGURE 4.35

FIGURE 4.36

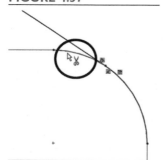

FIGURE 4.37

The result of the trimming operation is shown in Figure 4.36.

Repeat this process on the other arc/line intersection, as shown in Figure 4.37.

Add the two lines indicated in Figure 4.38 to close the sketch contour.

If the lines are horizontal and vertical, then the sketch should be fully defined.

Select the Extruded Boss/Base Tool from the Features group of the CommandManager. Extrude the rib as a Mid Plane Extrusion with a thickness of 0.125 inches, as shown in Figure 4.39.

The result is shown in **Figure 4.40**.

While the geometry of the rib appears to be satisfactory, there are two areas where the faces do not blend smoothly. These areas are the intersections between the rib and the fillets of the flange. The intersection at the lower fillet is shown in **Figure 4.41**; a similar area exists at the intersection with the upper fillet. We have a flat part (the rib) mating with a curved surface (the fillet). While the contour of the rib matches that of the fillet perfectly at the mid-surface of the rib, as the rib is extruded outward its surface is slightly higher than the fillet's. Although the difference is small, this type of mismatch can result in errors when exporting the geometry to a rapid prototyping system, tool path program, or finite element analysis program.

One solution to this mismatch would be to replace the extruded rib with one that is created by revolving the cross-section about the center axis of the flange through a small angle. However, this would result in a rib that is much thicker toward the outer edge of the flange. We could also try adding some small fillets to smooth the mismatched surfaces. A better approach is to use a revolved cut to remove the portions of the rib that extend above the fillets. Think of this operation as creating a cutting tool that precisely matches the profile of the mid-surface of the rib, and is swept around the center axis of the flange.

One way to match the rib profile is to open a new sketch and repeat the steps of identifying the silhouettes of the fillets and converting them to arcs, adding a line tangent to both arcs, and trimming away the excess portions, as we did previously. We cannot do this in the same location as the previous sketch, however, because the rib geometry has eliminated the silhouettes of the fillets. We could recreate the sketch steps in a new sketch in another location, say in the Right Plane, but it is easier to reuse the previous sketch. Another option would be to copy the previous sketch and edit it to create the cutter geometry, but the copied sketch would lose associativity with the rest of the flange geometry and would therefore not rebuild correctly if a change were made (for example, if the offset distance of Plane1 were to be changed).

FIGURE 4.38

FIGURE 4.39

FIGURE 4.40

FIGURE 4.41

Before adding on to the previous sketch, we need to make a small change to the definition of the rib. Since the sketch (Sketch4 in the FeatureManager) has only one closed contour, it was not necessary to select the contour when defining the rib (Extrude4). However, when we edit the sketch, we will be adding a second contour. Therefore, there will be an ambiguity in the definition of the rib unless we specify the contour to be used.

FIGURE 4.42

Select Extrude4 in the FeatureManager, and select Edit Feature from the pop-up toolbar, as shown in Figure 4.42. Click the double arrow to expand the box labeled "Selected Contours." Click in the box to highlight it, as shown in Figure 4.43. Click in the closed region of the sketch to select it, as shown in Figure 4.44. Click the check mark.

FIGURE 4.43

FIGURE 4.44

If the step above is not performed, then the addition to the sketch that follows will result in an error rebuilding the rib. The error can be fixed by editing the rib and selecting the sketch contour, but it is easier to preclude the error by selecting the contour before editing the sketch.

Click on the rib sketch (Sketch4) to select it, and select Edit Sketch from the pop-up toolbar, as shown in Figure 4.45. Switch to the Bottom View. Add the four horizontal and vertical lines indicated in Figure 4.46, and the two dimensions, which will fully define the sketch.

FIGURE 4.45

FIGURE 4.46

The 0.5-inch dimensions are arbitrary. However, by placing these dimensions, we are assured that the cutter's contour will always be larger than the flange, even if the flange dimensions change.

Rebuild the part. Click on Sketch4 again to select it, and choose the Revolved Cut Tool from the Features group of the Command-Manager. Click on the upper portion of the sketch to select it as the contour to be revolved, as shown in Figure 4.47.

FIGURE 4.47

We need to define the axis of the revolution. We could add a centerline to the sketch, but instead we will use the center axis of the flange. Every cylindrical feature has an associated axis, called a temporary axis. By default, these axes are not visible.

From the Heads-Up Toolbar, click the Hide/Show Items icon and click Temporary Axes, as shown in Figure 4.48. Click in the Revolve Parameters box in the PropertyManager to highlight it, and then click on the axis at the center of the flange, as shown in Figure 4.49. Click the check mark to complete the cut.

FIGURE 4.48

FIGURE 4.49

FIGURE 4.50

The rib and the fillets now interface smoothly, as shown in **Figure 4.50**. (Note that there is a tangent edge between the rib and fillet. If the display of tangent edges is turned off, then no edge will be visible. If the tangent edge is visible, select Options: System Options: Display/Selections and check Removed for Part/Assembly tangent edge display, as we did in Chapter 1.)

FIGURE 4.51

Note that Sketch4 is shown in the FeatureManager under both the rib (Extrude4) and the cut (Cut-Revolve1). The hand under each sketch icon, as shown in **Figure 4.51**, indicates that the sketch is shared with another feature.

We will now add draft to the rib. Unlike the base feature and the first extruded boss, where the draft was specified as part of the extrusion step, the necessary draft on the rib is not in the same direction as the extrusion. Therefore, draft must be added as a secondary operation.

FIGURE 4.52

Select the Draft Tool from the Features group of the CommandManager, as shown in Figure 4.52. For the Neutral Plane, select the flat surface shown in Figure 4.53 (make sure not to select the curved fillet surface).

Note the arrow pointing away from the flange body. This is the direction in which the feature will get smaller as the draft is applied (i.e., the direction in which a mold half would be pulled). Since this is the correct direction for the draft, we do not need to change it.

FIGURE 4.53

For the Faces to Draft, click on the surface shown in Figure 4.54. Rotate the flange and select the other side of the rib as well. Set the draft angle to 2 degrees, as shown in Figure 4.55. Click the check mark to apply the draft.

The draft can be seen clearly from the Front View, as shown in Figure 4.56. The top of the flange is wider toward the outer perimeter of the flange, since the surface is farther away from the neutral surface. The width of the base of the flange is constant.

FIGURE 4.54

FIGURE 4.56

FIGURE 4.55

We will now create a pattern to add the other ribs.

Select the Circular Pattern Tool from the pull-down menu under the Linear Pattern Tool. Expand the fly-out FeatureManager by clicking the + sign beside the part name. For the features to pattern select the rib (Extrude 4), the cut (Cut-Extrude1), and the draft, as shown in Figure 4.57. Click in the Pattern Axis

FIGURE 4.57

FIGURE 4.58

box to highlight it, and select the temporary axis at the center of the part, as
shown in Figure 4.58. Set the number of
instances to 6, and click the check mark.
Turn off the display of the temporary
axes from the Hide/Show Items Tool of the
Heads-Up Toolbar.

The rib pattern is shown in **Figure 4.59**.

The holes will now be added.

Select the face shown in Figure 4.60, and change to the Normal To View.

FIGURE 4.59

FIGURE 4.60

**Select the Circle Tool from the Sketch group of the CommandManager. Draw a cir-
cle centered at the origin. Check the "For construction" box in the Property-
Manager. Select the Smart Dimension Tool, and dimension the circle diameter as
4.5 inches.**

**Select the Centerline Tool
and draw two centerlines,
both originating from the
origin. One of the centerlines should be horizontal
and the other diagonal, as
shown in Figure 4.61. Select
the Smart Dimension Tool
and add a 30-degree angular
dimension between the
centerlines.**

FIGURE 4.61

Select the Circle Tool. Move the cursor to the intersection of the construction circle and the diagonal centerline, so that the intersection icon appears, as shown in Figure 4.62.

Drag out a circle. Select the Smart Dimension Tool and add a 3/8-inch diameter dimension to the circle, as shown in Figure 4.63.

FIGURE 4.62   **FIGURE 4.63**

Select the Extruded Cut Tool from the Features group of the CommandManager. Set the type as Through All, and click the check mark to create the first hole.

Select the Circular Pattern Tool. Create a six-hole pattern.

The finished part is shown in Figure 4.64.

In the next section, we will add equations to link the rib and hole patterns. This will be easier to do if we rename these features.

In the FeatureManager, rename the extrusion defining the first rib "Rib," the first circular pattern "Rib Pattern," the first bolt hole "Bolt Hole," and the second circular pattern "Hole Pattern," as shown in Figure 4.65. Save the part.

FIGURE 4.64   **FIGURE 4.65**

4.2 Creation of Parametric Equations

The use of parametric equations embedded in models to control dimensions is a powerful engineering tool. These equations can be used to embed design intelligence directly into solid models. In this section, the flange created in Section 4.1 will be modified to include parametric equations relating the hole pattern to the rib pattern.

With the flange model open, select Tools: Equations from the main menu, as shown in Figure 4.66.

The Equations dialog box will appear.

In the Equations dialog box, click the Add button, as shown in Figure 4.67.

FIGURE 4.66

FIGURE 4.67

The Add Equation dialog box will appear, as shown in Figure 4.68. This dialog box allows for equations relating model parameters to one another to be added to the model.

Double-click on the Hole Pattern in the Feature-Manager.

This will display the parameters (6 holes, 360° spacing) in the model window.

Click on the "6" parameter (denoting the number of instances) in the model window, as shown in Figure 4.69.

FIGURE 4.68

FIGURE 4.69

Note that the symbol denoting this parameter ("D1@Hole Pattern") now appears in the Add Equation dialog box. We can now establish a parametric relationship driving this parameter (the number of holes in the pattern) with another parameter in the model. When considering our design intent, the number of holes in the pattern is not necessarily an independent design choice; most likely, we simply want to ensure that there is one hole centered between each pair of stiffening ribs. Therefore, we will establish a parametric equation tying the number of holes in the pattern directly to the number of ribs in the Rib Pattern.

Click on the equal sign as shown in Figure 4.70.

FIGURE 4.70

Since the parameter representing the number of holes in the Hole Pattern is on the left side of this equal sign, the number of holes is a driven parameter; whatever expression is entered on the right side of the equal sign will determine its value. Note that since this value is now a driven parameter, it cannot be directly modified anymore.

The right side of the equation will now be added.

FIGURE 4.71

Double-click on the Rib Pattern to display the parameters associated with it. Click on the "6," as shown in Figure 4.71, to complete the equation.

The equation is shown in **Figure 4.72**.

Select OK to add the equation. The Equations dialog box now shows the equation and the value that is calculated for the equation (6), as shown in Figure 4.73. Click OK to close the dialog box.

FIGURE 4.72

FIGURE 4.73

This equation establishes a parametric relation that ties the number of holes in the hole pattern directly to the number of ribs in the rib pattern. We will test the equation at this point.

Double-click the Rib Pattern to display its parameters, and double-click on the value "6" (the number of ribs in the pattern) and change it to "3." Click the check mark, and then click the Rebuild Tool to rebuild the model.

FIGURE 4.74

The model will be rebuilt, with the Rib Pattern modified and the driven Hole Pattern modified as well, as shown in Figure 4.74.

While the number of holes is correctly tied to the number of ribs, our design intent may not be satisfied by this model. Note that the angular location of the holes was set at 30 degrees from the center of a rib; this centered the hole between two ribs when there were six ribs in the pattern, but it no longer provides for centering of the holes when the number of ribs is modified. We will create a new parametric equation to establish the relationship required by our design intent; the equation will drive the angular dimension of the holes so that they are centered between the ribs.

From the main menu, select Tools: Equations. In the Equations dialog box, click Add. Double-click the Bolt Hole in the FeatureManager to display the dimensions associated with the bolt hole.

The angular spacing between ribs is 360 degrees divided by the number of ribs. The first hole is located at one-half of this value away from the first rib. Therefore, the equation required to set the angular dimension in degrees to the desired value will be:

$$\text{Angular Dimension} = (1/2)\ (360/\text{Number of Ribs}) \text{ or}$$

$$\text{Angular Dimension} = 180/\text{Number of Ribs}$$

FIGURE 4.75

With the hole dimensions still displayed, select the angular dimension locating the hole, as shown in Figure 4.75. In the Add Equation dialog box, click the "=" sign. Type the number 180, and click the "/" sign.

Double-click on the Rib Pattern in the Feature-Manager to display the associated parameters. Select the parameter that represents the number of ribs by clicking on the number 3 in the model window.

FIGURE 4.76

The new equation should be as shown in Figure 4.76.

Click OK to close the Add Equation dialog box.

The Equations box confirms that the dimension will equal 60 degrees with the current number of ribs (3), as shown in **Figure 4.77**.

Click OK to close the Equations box and then rebuild the model.

The result is shown in **Figure 4.78**.

Change the number of ribs to other values and check to see that the number and locations of the holes change in a consistent manner.

Another example, with five ribs and holes, is shown in **Figure 4.79**.

Note that there is a difference between *driving* and *driven* parameters. In this example, the number of ribs is a free design choice, and is therefore a driving parameter. As such, it appears only on the right side of the equal sign in parametric equations. Conversely, the number of holes and the angular hole location are driven parameters; they are determined by the choice of the number of ribs, and appear on the left side of the equal sign in parametric equations. Since their values are set by the values established by the driving parameters, driven parameters cannot be modified directly in the model.

Perform the following demonstration to verify this.

FIGURE 4.77

FIGURE 4.78

FIGURE 4.79

Double-click on the Hole Pattern in the FeatureManager to display the associated parameters in the model window. Double-click on number of holes in the pattern, to try to change the value.

FIGURE 4.80

The "Σ" symbol displayed beside the value (see **Figure 4.80**) indicates that the dimension is controlled by an equation and cannot be changed.

Click the check mark to close the message box. Reset the number of ribs and holes to six, rebuild the model, and save the part file.

4.3 Modeling Tutorial: Cap Screw with Design Table

Note: This tutorial requires Microsoft Excel to be installed on your computer.

In this section, we will create a family of similar parts. Many parts are defined this way, especially common parts such as fasteners, washers, seal rings, and so on. Rather than creating separate model files and drawings for every different part, a single part with multiple configurations is made. A single drawing can be made to define the parameters of all of the different configurations. The specifications of the dimensions that define each configuration are contained in a spreadsheet called a *design table*.

Open a new part. Select the Right Plane from the FeatureManager, and select the Circle Tool from the Sketch group of the CommandManager. Draw a circle centered at the origin. Select the Smart Dimension Tool, and dimension the diameter of the circle as 0.50 inches, as shown in Figure 4.81.

FIGURE 4.81

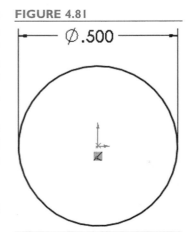

Select the Extruded Boss/Base Tool from the Features group of the CommandManager, and extrude the circle 1.50 inches, in the direction shown in Figure 4.82.

Select the face shown in Figure 4.83, and select the Normal To View.

Select the Circle Tool from the Sketch group of the CommandManager. Draw a circle centered at the origin. Select the Smart Dimension Tool and dimension the diameter of the circle as 0.75 inches, as shown in Figure 4.84.

FIGURE 4.82

Direction of Extrusion

FIGURE 4.83

FIGURE 4.84

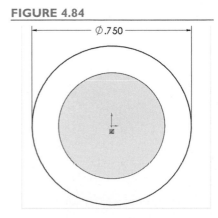

Select the **Extruded Boss/Base Tool** from the Features group of the Command-Manager, and extrude the circle 0.50 inches to form the head of the screw, as shown in Figure 4.85.

Select the top of the screw head, as shown in Figure 4.86. Choose the **Normal To View.**

FIGURE 4.85

Direction of Extrusion

FIGURE 4.86

Select the **Polygon Tool** from the Sketch group of the CommandManager, as shown in Figure 4.87. Drag out a polygon from the origin, as shown in Figure 4.88. By default, the number of sides is six (a hexagon).

Click the check mark in the Property-Manager to turn off the **Polygon Tool,** and click on one of the sides of the hexagon to select it. In the Property-Manager, add a **Horizontal** relation to the side, as shown in Figure 4.89.

Select the **Smart Dimension Tool** from the Sketch group of the CommandManager, and add a 0.375-inch dimension between two opposite sides of the hexagon, as shown in Figure 4.90.

The sketch is now fully defined.

FIGURE 4.87

FIGURE 4.88

FIGURE 4.89

FIGURE 4.90

FIGURE 4.91

Select the Extruded Cut Tool from the Features group of the CommandManager, and extrude a cut 0.245 inches deep to form the hex socket in the head, as shown in Figure 4.91.

Screw threads are rarely modeled as solid features, since they can be cumbersome to create and can significantly slow down the performance of your computer. Also, since they are standardized, the thread profiles are not required in order to specify them on a drawing. Rather, *cosmetic threads*, graphical features used to show the threaded regions, are much more commonly used. Cosmetic threads can be added to any cylindrical feature.

Select the edge where the threads will begin, as shown in Figure 4.92.

FIGURE 4.92

Select Insert: Annotations: Cosmetic Thread from the main menu, as shown in Figure 4.93.

FIGURE 4.93

Set the thread length as 1.00 inch, and the minor diameter as 0.40 inches, as shown in Figure 4.94. **Click the check mark to add the thread. Switch to the Front View. The resulting thread display is shown in** Figure 4.95 **(shown in wireframe mode for clarity).**

FIGURE 4.94

FIGURE 4.95

The minor diameter is the diameter at the "root" of the thread. The actual minor diameter value of a UNC (unified series coarse) 1/2-inch thread is 0.408 inches. Since the minor diameter of the cosmetic thread is only for display purposes, using an approximate value is acceptable.

If preferred, the cosmetic thread can be displayed with a shaded thread pattern on the surface rather than with the dashed lines shown in **Figure 4.95**. If you want to change the display mode of the cosmetic thread, choose the Options Tool, and under the Document Properties tab, click on Detailing. Click on the "Shaded cosmetic threads" box, as shown in **Figure 4.96**. The resulting display is shown in **Figure 4.97**. Note that some lim-

FIGURE 4.96

ited-release versions of the software may not support the display of shaded threads.

The definition of the thread appears in the FeatureManager, attached to the cylindrical feature that is to be threaded, as shown in **Figure 4.98**.

FIGURE 4.97   **FIGURE 4.98**

Save the part file as "Cap Screw."

We have now defined one configuration of the cap screw. To define more configurations, we are going to create a design table.

We will need to access all of the dimensions used to create the part.

To show the dimensions, right-click on Annotations in the FeatureManager, and select Show Feature Dimensions, as shown in Figure 4.99.

FIGURE 4.99

Move the dimensions around the screen so that they are all easily visible, as shown in Figure 4.100.

The Isometric View is used here. Since the dimension text is aligned with the plane in which it was created, the Isometric View allows for the best display of all dimensions.

If desired, the display of dimensions can be changed so that the text is always oriented relative to the screen by selecting Tools: Options: System Options: Display/ Selection and checking the box labeled "Display dimensions flat to screen."

From the main menu, select Insert: Tables: Design Table, as shown in Figure 4.101.

FIGURE 4.100

FIGURE 4.101

FIGURE 4.102

In the PropertyManager, leave the Source option as Auto-create. In the Edit Control box, choose the second option, as shown in Figure 4.102, so that dimensions entered in the table cannot be changed outside of the table. Clear all of the options for adding new rows and columns, and click the check mark to begin creating the table.

A dialog box will open, prompting you to select the dimensions that will be included in the table, as shown in Figure 4.103. Select all of them by clicking on the first and, while holding the Shift key, clicking on the last. Click OK.

FIGURE 4.103

FIGURE 4.104

A window containing a Microsoft Excel spreadsheet will open, as shown in **Figure 4.104**. Note that the Excel tools are available at the top of the screen whenever the table is open. Be careful not to click in the white space of the graphics window while the table is open; doing so will close the table. If you close the table accidentally, you can re-open it by clicking on the ConfigurationManager tab above the FeatureManager, right-clicking the table name in the ConfigurationManager, and selecting Edit Table.

Note that row 2 of the table contains the SolidWorks name of each dimension, and row 3 contains the current values of the dimensions as the "default" configuration. The only dimension that we will need to add manually is the thread length, since dimensions associated with cosmetic threads are not automatically added to design tables.

FIGURE 4.105

FIGURE 4.106

Click in cell H2 to select it, as shown in Figure 4.105. Double-click the dimension defining the length of the cosmetic threads, as shown in Figure 4.106.

The name of the dimension will be placed in cell H2. The dimension's current value (1 inch) will be placed in cell H3, as shown in **Figure 4.107**.

FIGURE 4.107

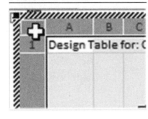

FIGURE 4.108

To use a larger font size, click the blank area in the upper left corner of the table, as shown in Figure 4.108. This selects all of the cells in the table. Choose a new font size (14-16 point is a good choice) from the Home group of the Excel tool ribbon. To adjust the width of each column, move the cursor to the right boundary of the column's header, as shown in Figure 4.109, and click and drag to change the width. Change the column widths so that all of the digits of the dimensions are displayed.

FIGURE 4.109

Width: 5.00 (40 pixels)

D	E	F	G	H	I
: Cap Screw					
				d1	

FIGURE 4.110

Change the name of the default configuration to "Part 101" in cell A3. Enter two more configurations and their dimensions in the Table, as shown in Figure 4.110.

When the table is closed, the program will search column A (below row 1) for configuration names, much like a lookup table in Excel.

	A	B	C	D	E	F	G	H
1	Design Table for: Cap Screw							
2		D1@Sketch1	D1@Extrude1	D1@Sketch2	D1@Extrude2	D1@Sketch3	D1@Extrude3	D1@Cosmetic Thread1
3	Part 101	0.5	1.5	0.75	0.5	0.375	0.245	1
4	Part 102	0.75	2	1.125	0.75	0.625	0.37	1.5
5	Part 103	1	3	1.5	1	0.75	0.495	2

Click in the white space in the modeling window, outside of the spreadsheet window. This will cause the design table to close, and a message will be displayed that indicates the new configurations have been created, as shown in Figure 4.111. Click OK.

FIGURE 4.111

Note that the part has not changed on the screen. That is because the default configuration is the one with the dimensions used to model the part originally. To view the new configurations created from the design table, we need to use the ConfigurationManager.

At the top of the FeatureManager, there are tabs corresponding to the FeatureManager, PropertyManager, and ConfigurationManager. (There could be other tabs as well, if certain add-ins are present.)

FIGURE 4.112

FIGURE 4.113

Click on the icon representing the ConfigurationManager, as shown in Figure 4.112.

In the ConfigurationManager, double-click on Part 103, as shown in Figure 4.113.

FIGURE 4.114

The part is rebuilt to the dimensions specified in the design table for Part 103, as shown in **Figure 4.114**. Dimensions controlled by the design table may appear in a different color on your screen. Note that when you move a dimension, you will receive a message that the dimension's value cannot be edited, as shown in **Figure 4.115**.

FIGURE 4.115

One dimension that was left unchanged is the minor thread diameter of the cosmetic thread. Since this is not a dimension used to define the part, we did not include it in the design table. However, we would like to have the display of the cosmetic thread look reasonable on the screen. We can add an equation to control this dimension.

From the main menu, select Tools: Equations.

Click Add to create a new equation. Click on the dimension representing the minor thread diameter (0.400). In the equation box, type "=0.8*" after the name of the minor diameter dimension ("D2@Cosmetic Thread1") and click on the dimension representing the diameter of the shank (φ1.00). The equation should appear as shown in Figure 4.116.

FIGURE 4.116

FIGURE 4.117

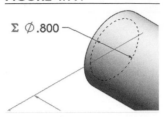

Click OK to add this equation, and OK again to close the Equation box.

Click the Rebuild Tool.

The updated thread diameter is shown in **Figure 4.117**.

FIGURE 4.118

Click on the FeatureManager tab to return to the FeatureManager. Right-click Annotations in the Feature-Manager, and select Show Feature Dimensions, as shown in Figure 4.118, to turn off the display of the dimensions on the screen. Save the part file.

4.4   Incorporating a Design Table in a Drawing

We now will make a single drawing that details all three configurations of the cap screw.

Open a new drawing. Choose an A-size landscape sheet size, and either uncheck the "Display sheet format" box for a plain sheet, or select the sheet format that you created in Chapter 2.

If the Model View command does not start automatically, then select the Model View Tool from the Drawings group of the CommandManager. If desired, check the box labeled "Start command when creating new drawing," as shown in Figure 4.119. Click on the Cap Screw in the Open documents box, or browse to find it. Click the Next arrow in the PropertyManager.

Select "Multiple views," and select the Front and Right Views. Choose the wireframe display style with the hidden lines visible, as shown in Figure 4.120.

FIGURE 4.119

FIGURE 4.120

Click the check mark to create the views, which are shown in Figure 4.121.

FIGURE 4.121

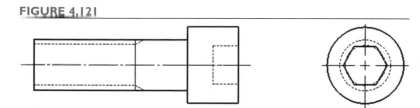

The model views that are created reflect the configuration that is current when the drawing is created. The configuration used for any model view can be changed by right-clicking on that view and selecting Properties from the menu that appears. For this tutorial, the configuration used is not important.

With one of the drawing views selected, select Design Table from the Table Tool in the Annotation group of the CommandManager, as shown in Figure 4.122.

FIGURE 4.122

The design table is inserted into the drawing, as shown in **Figure 4.123**. The design table can be moved on the drawing sheet by clicking and dragging it, but the format of the table is not what we want. There are blank rows and columns visible, dimensions are shown to only the number of decimal places we input, and so on. We will now edit the table to improve its appearance.

FIGURE 4.123

Design Table for: Cap Screw

	D1@Sketch1	D1@Extrude1	D1@Sketch2	D1@Extrude2	D1@Sketch3	D1@Extrude3	D1@Cosmetic Thread1
Part 101	0.5	1.5	0.75	0.5	0.375	0.245	1
Part 102	0.75	2	1.125	0.75	0.635	0.37	1.5
Part 103	1	3	1.5	1	0.75	0.495	2

FIGURE 4.124

Double-click on the design table, and you will be taken back to the part screen, with the spreadsheet open in a window, as shown in Figure 4.124.

Remember that the design table is a Microsoft Excel spreadsheet, and we will be editing it the same way we would edit any other Excel spreadsheet. To begin, the first two rows do not need to be displayed on the drawing. We cannot delete them, since they contain information necessary to the design table, but we can hide them.

Click and hold the mouse key on the number 1 on the left side of the spreadsheet, drag the cursor down onto the 2, and release. Rows 1 and 2 will be highlighted, as shown in Figure 4.125. Right-click, and select Hide.

Select Columns B–H. Click the Center icon on the Excel menu, as shown in Figure 4.126.

FIGURE 4.125

FIGURE 4.126

We will now add headers to our table. We hid the row containing the SolidWorks names of the dimensions, and we will now add a row that contains more descriptive names.

Right-click on the "3" at the left side of the table, and select Insert, as shown in Figure 4.127. In the new row that is inserted, type "Diameter" in the cell in column B, as shown in Figure 4.128.

FIGURE 4.127

FIGURE 4.128

Note that the text is rotated 90 degrees. By default, the SolidWorks part names were rotated to save space. However, for displaying on our drawing, we would prefer that the names are not rotated.

Click on the "3" on the left side of the table to select that row. Click the arrow beside the Orientation Tool in the Home group of the Excel ribbon, as shown in Figure 4.129. The Rotate Text Up tool is selected; click to de-select it, as shown in Figure 4.130.

FIGURE 4.129

FIGURE 4.130

Enter the other names, as shown in Figure 4.131. Adjust the column widths so that the labels are displayed completely. To show the longer names on two lines, highlight the desired cells and choose the Wrap Text Tool, as shown in Figure 4.132.

Be sure to leave cell A3 blank. Recall that column A is reserved for configuration names.

FIGURE 4.131

	A	B	C	D	E	F	G	H	I
3		Diameter	Length	Head Diameter	Head Height	Hex Width	Hex Depth	Thread Length	
4	Part 101	0.5	1.5	0.75	0.5	0.375	0.245	1	
5	Part 102	0.75	2	1.125	0.75	0.625	0.37	1.5	
6	Part 103	1	3	1.5	1	0.75	0.495	2	
7									
8									
9									

Sheet1

Highlight all of the cells containing numerical values. Click the Increase Decimal Tool, shown in Figure 4.133, until all values are shown to three decimal places.

FIGURE 4.132

FIGURE 4.133

Click and drag the "handle" at the lower right corner of the table until all blank rows and columns are hidden, as shown in Figure 4.134. Click in the white space of the graphics area to close the table. From the menu, select Windows and choose the name of the drawing.

FIGURE 4.134

G	H
Hex Depth	Thread Length
0.245	1.000
0.370	1.500
0.495	2.000

FIGURE 4.135

	Diameter	Length	Head Diameter	Head Height	Hex Width	Hex Depth	Thread Length
Part 101	0.500	1.500	0.750	0.500	0.375	0.245	1.000
Part 102	0.750	2.000	1.125	0.750	0.625	0.370	1.500
Part 103	1.000	3.000	1.500	1.000	0.750	0.495	2.000

Click on the Rebuild Tool to update the design table in the drawing, which now appears as shown in Figure 4.135.

Before adding dimensions to the drawing, we need to add a view. Since the depth of the hex cavity needs to be dimensioned, we will add a section view of the head. This will let us avoid dimensioning a hidden feature. It is good practice to refrain from using hidden lines for dimensioning.

FIGURE 4.136

Zoom in on the head of the screw in the Front View. Choose the Section View Tool from the View Layout group of the Command-Manager. Hold the cursor momentarily at the midpoint of the top edge of the head to "wake up" this feature, and then move the cursor to the right, as shown in Figure 4.136. The coincident relation icon will appear, indicating that the cursor is aligned horizontally with the midpoint of the top edge of the head. Click and drag a horizontal line through the head, as shown in Figure 4.137. Note the coincident and horizontal relation icons. When you click to complete the line, you will see a message that a partial section view will be created, as shown in Figure 4.138. Click Yes to complete the view.

FIGURE 4.137

Drag the section view away from the Front View and click to place it. Right-click, and choose Alignment: Break Alignment, as shown in Figure 4.139.

FIGURE 4.138

FIGURE 4.139

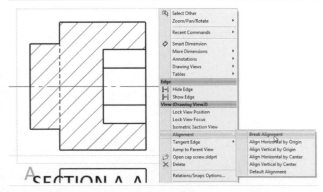

You will now be able to click and drag the section view to any location on the sheet, as shown in Figure 4.140.

With the section view selected, choose **Hidden Lines Removed** from the PropertyManager, as shown in Figure 4.141.

We will now add dimensions to the drawing. Since we want most of the dimensions to appear in the Front View, we will import dimensions into that view first.

Select the Front View. Select the Model Items Tool from the Annotations group of the CommandManager. In the PropertyManager, select Entire Model as the source, as shown in Figure 4.142. Make sure that the box labeled "Import items into all views" is unchecked. Click the check mark to apply the dimensions.

FIGURE 4.140

FIGURE 4.141

FIGURE 4.142

FIGURE 4.143

SECTION A-A

Repeat for the Right View and then for the section view, so that all dimensions are imported, as shown in Figure 4.143.

The dimensions shown are for the Part 103 configuration, which was the configuration selected when the part file was saved. If you saved the part with another configuration selected, then the dimensions will be different. This is not a problem, as we will be replacing the numerical values with the dimension names.

The dimensions defined by the design table may be shown in a different color than black. (The default color is magenta.) While this may be desirable in the part file, for the drawing we would prefer that all dimensions be displayed in black.

Select the Options Tool. Under the System Options tab, select Colors and scroll down the list of entities to find "Dimensions, Controlled by Design Table," as shown in Figure 4.144. Select Edit, and set the color to black.

FIGURE 4.144

Move the dimensions and/or views on the screen so that they are all visible.

The dimensions displayed are those for one configuration. To make this drawing one that defines the dimensions for all configurations, we need to display the dimension names in the drawing views.

Click on the dimension defining the length, as shown in Figure 4.145. **In the PropertyManager, replace the "<DIM>" in the Dimension Text box and type in "Length," as shown in** Figure 4.146. **A message will appear, warning you that any tolerances specified for this dimension will not be displayed (see** Figure 4.147). **Check the "Don't ask me again" box, and click Yes.**

FIGURE 4.145

FIGURE 4.145

FIGURE 4.146

FIGURE 4.147

Confirm dimension value text override

Overriding the dimension value text <DIM> disables tolerance display. Do you want to continue?

Yes   No

☑ Don't ask me again

Repeat for the other dimensions. Add a note calling out UNC-2A threads (UNC = Unified Standard Coarse series, 2 = tolerance level, A = external threads) and a note specifying inches as the dimensions. The completed drawing is shown in Figure 4.148.

FIGURE 4.148

	Diameter	Length	Head Diameter	Head Height	Hex Width	Hex Depth	Thread Length
Part 101	0.500	1.500	0.750	0.500	0.375	0.245	1.000
Part 102	0.750	2.000	1.125	0.750	0.625	0.370	1.500
Part 103	1.000	3.000	1.500	1.000	0.750	0.495	2.000

All dimensions are inches

SECTION A-A

Hex Depth

Length
Thread Length
Diameter
UNC-2A Threads
Head Height
Head Diameter
Hex Width

Introduction to Solid Modeling Using SolidWorks 2009

Chapter 4 - Cap Screw

PROBLEMS

P4.1   Consider the flange model developed in Chapter 1. Create a "blank" for this part (without the holes), as shown in **Figure P4.1A**, using the following procedure:

a.   Sketch a 5.5-inch diameter circle on the Top Plane, and extrude it upward 2.25 inches to create a solid cylinder (**Figure P4.1B**).

b.   In the Front Plane, sketch a "cutting tool" to create the flange "blank" from the solid (**Figure P4.1C**).

c.   Use the Revolved Cut command to create the blank.

FIGURE P4.1A   **FIGURE P4.1B**

FIGURE P4.1C

P4.2 Create a solid box, as shown in **Figure P4.2A**. Using equations, set the depth $d = 2w$, and the height $h = 3w$, where the width w is the driving dimension. Show the box for a few different values of the driving dimension, as shown in **Figure P4.2B**.

FIGURE P4.2A

FIGURE P4.2B

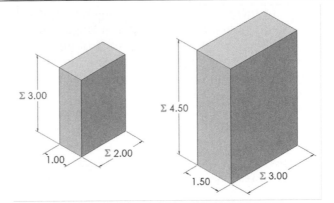

P4.3 Model the bracket shown in **Figure P4.3A**. Use only the dimensions shown in **Figure P4.3B**; use relations and symmetry as required so that these dimensions completely define the part. (The fillet radius, 0.125 inches, is the same for the three fillets.)

Add equations to the drawing so that these relationships exist between the dimensions:

1. Height = 0.40 (Width)
2. Leg Depth = 1/2 (Height)
3. Hole Spacing = Width minus 2 inches
4. Hole Location = 1/2 (Height minus 0.25 inches)
5. Slot Width = 1/2 (Hole Spacing)
6. Slot Location = 1/2 (Leg Depth)

Check to see that the equations work for Width values from 3 to 8 inches.

Note: The slot can be created using the Slot Tool from the Sketch group of the Command-Manager.

FIGURE P4.3A

FIGURE P4.3B

Introduction to Solid Modeling Using SolidWorks 2009

Chapter 4 - Bracket

P4.4   In this exercise, you will model a part in which an integer design parameter (number of holes) is controlled by an equation.

a.   Model the part shown in **Figure P4.4A**. Use a linear pattern to place the second hole. Show all of the dimensions by right-clicking on Annotations in the FeatureManager and clicking on "Show Feature Dimensions."

b.   Add an equation so that the number of holes is equal to the length of the part (4 inches in the current configuration) divided by two. Therefore, there will be one hole for every 2 inches of length. Show only the dimensions shown in **Figure P4.4B** by right-clicking on each of the other dimensions and selecting "Hide."

c.   Change the length of the part to 5 inches, as shown in **Figure P4.4C**. Note that the program has rounded the value of the equation (2.5) to the closest integer value (3).

d.   Add a second equation to change the value of the dimension specifying the location of the first hole so that the holes will be centered on the part, as shown in **Figure P4.4D**.

Experiment with several values of length to show that the equations produce the desired results, as shown in **Figures P4.4E** and **P4.4F**.

FIGURE P4.4A

FIGURE P4.4B

FIGURE P4.4C

FIGURE P4.4D

FIGURE P4.4E

FIGURE P4.4F

P4.5 Revisit the perforated board created in Problem P3.5. Add equations to resize the overall board dimensions based on the number of holes needed in the board (using the same hole size and spacing used in the original problem). Show that the equations work by generating new boards for the following cases:
a. A 70 × 50 grid of holes
b. A 80 × 30 grid of holes
c. A 45 × 15 grid of holes

P4.6 Revisit the hacksaw blade created in Problem P3.6. Add equations to calculate the proper tooth spacing and number of teeth in the linear pattern based on the blade length (the 12-inch dimension in Figure P3.6C) and the tooth length (the 0.050-inch dimension in Figure P3.6D). Show that the equation works by creating blades for the following three cases:
a. 18-inch length, 0.050-inch tooth length
b. 15-inch blade length, 0.060-inch tooth length
c. 15.5-inch length, 0.060 inch-tooth length

P4.7 Consider the heat sink from Problem P3.9, which is shown in **Figure P4.7A**. Make a copy of this part, and add equations as follows (refer to **Figure P4.7B** for the dimension names):
1. The depth = one-half of the width plus 10 mm.
2. The fin height = one-third of the width.
3. The number of fins = the width divided by 6 mm.
4. The offset distance to the first fin is such that the fins are placed symmetrically of the part.

Show that the equations work correctly for width values from 60 to 180 mm.

FIGURE P4.7A

FIGURE P4.7B

P4.8   Use a design table to create the W18 series of wide-flange shape beams, according to **Figure P4.8** and **Table P4.8**. The model of the beam should be extruded to 36 inches in all configurations.

FIGURE P4.8

Table P4.8

Designation	Depth d	Flange width b	Flange thickness T	Web thickness t
W18 x 106	18.73	11.200	0.940	0.590
W18 x 76	18.21	11.035	0.680	0.425
W18 x 50	17.99	7.495	0.570	0.355
W18 x 35	17.70	6.000	0.425	0.300

Note: All dimensions in inches.

P4.9   Create a multiconfiguration drawing of the model created in **Figure P4.8**.

P4.10   Make a copy of the flange created in Chapter 1, which is shown in **Figure P4.10A** with some of the dimensions hidden. Create three different configurations of the flange (the flange as created in Chapter 1 will be the first configuration), using both equations and a design table, as detailed below.

FIGURE P4.10A

     a.   Add two equations:
         (1)   The boss diameter (2.75 in.) is equal to one-half of the flange diameter (5.50 in.).
         (2)   The center hole diameter (1.50 in.) is equal to the boss diameter minus 1.25 inches.
     b.   Create a design table to define the dimensions of two additional configurations, as specified in **Table P4.10**.
     c.   Make a 2-D drawing showing the three configurations, with only the dimensions that change shown, as in **Figure P4.10B**. This type of drawing is used often in product literature and catalogs to illustrate the relative sizes of different parts.

Insert three Front Views and three Top Views in the drawing. Right-click on each view and select Properties, and select the appropriate named configuration for each view. Hide unwanted dimensions by selecting View: Hide/Show Annotations from the main menu

Table P4.10

Flange	Diameter	Height	Bolt circle diameter	Number of bolt holes
Part 1	5.5	2.25	4.25	4
Part 2	7.0	3.00	5.25	6
Part 3	8.0	3.50	6.25	8

Note: All dimensions in inches.

and selecting the dimensions to be hidden. Press the Esc key to return to the normal drawing mode.

To precisely align the views, select a view, right-click, and choose Align: Align Horizontal by Origin, and select another view to align to. Repeat until all views are aligned.

FIGURE P4.10B

Introduction to Solid Modeling
Using SolidWorks 2009

Chapter 4 - Flange

P4.11 An O-ring is an elastomeric seal in the shape of a torus, as shown in **Figure P4.11A**. Standard sizes of O-rings used in the United States are defined by a Society of Automotive Engineers (SAE) specification. Create a solid model of an O-Ring, with the six configurations as detailed in **Table P4.11**. (Note that while it is normally good practice to dimension to the centers of circular features rather to their edges, O-rings are defined by their inner diameters. This allows a design engineer to determine how much the seal will have to stretch to fit into a groove of a specified size.) Make a multi-configuration drawing, as shown in **Figure P4.11B**.

FIGURE P4.11A

Table P4.11

Part Number	Inner Diameter	Thickness
125	1.299 ± .020	.103 ± .003
142	2.362 ± .020	.103 ± .003
235	3.109 ± .024	.139 ± .004
254	5.484 ± .035	.139 ± .004
350	4.600 ± .030	.210 ± .005
362	6.225 ± .035	.210 ± .005

When adding the inner diameter dimension to the sketch defining the part geometry, note that the Smart Dimension Tool attempts to define the dimension to the center of the circle representing the O-ring cross-section. To dimension to the edge of the circle, use the Point Tool to add a point to a quadrant point of the circle, and dimension to that point.

While the tolerance values cannot be used to change values within the part (even if you define the tolerance associated with a dimension on the part file, only the nominal value appears in the design table), they can be added in separate columns in the design table. To add the plus/minus symbol, select Insert: Symbol from the Excel tools.

FIGURE P4.11B

	Inner Diameter	Thickness	Diameter Tolerance	Thickness Tolerance
125	1.299	0.103	±.020	±.003
142	2.362	0.103	±.020	±.003
235	3.109	0.139	±.024	±.004
254	5.484	0.139	±.035	±.004
350	4.600	0.210	±.030	±.005
362	6.225	0.210	±.035	±.005

Dimensions are inches
Material = Buna-N Rubber

Thickness

Inner Diameter

SECTION A-A

Introduction to Solid Modeling Using SolidWorks 2009

Chapter 4 - O-Ring

CHAPTER 5

Advanced Concepts in Part Modeling

Introduction

The parts that we have made so far have been made primarily with extruded and revolved bases, bosses, and cuts. In this chapter, we will introduce several other tools for creating and modifying parts, including the *Loft*, *Sweep*, and *Shell* Tools.

5.1 A Lofted and Shelled Part

In this exercise, we will construct the business card holder shown in **Figure 5.1**. Note that the top of the part is rounded, while the bottom of the part is rectangular. These dissimilar shapes will be joined into a solid with the Loft Tool. Also, notice that the part is not solid, but rather is hollow underneath, as the view in **Figure 5.2** shows. The Shell Tool allows this type of construction to be easily modeled.

FIGURE 5.1

FIGURE 5.2

Chapter Objectives

In this chapter, you will:

- create a lofted feature from multiple sketches,

- use the Shell Tool to create a thin-wall part,

- learn how to change the order in which features are created and modified,

- create raised letters on a part, and

- create a feature by sweeping a cross-section around a planar or 3-D path.

157

Open a new part. Select the Top Plane. Select the Center Rectangle Tool from the Sketch group of the CommandManager. Drag out a rectangle, centered at the origin.

FIGURE 5.3

Select the Smart Dimension Tool, and dimension the rectangle to be 5.50 inches by 2.00 inches. The resulting sketch is shown in Figure 5.3.

In previous tutorials, upon completing a sketch we have then used a Features tool to convert the sketch into a 3-D object. However, a lofted feature requires at least two sketches. Therefore, we will close this sketch and begin the second sketch.

Close the sketch by clicking on the icon indicated in Figure 5.4, in the upper-right corner of the screen.

FIGURE 5.4

You can also close a sketch by choosing Exit Sketch from the Sketch group of the CommandManager.

The second sketch, which will define the top of the part, will be created in a new plane.

FIGURE 5.5

Select the Top Plane from the FeatureManager. Click on the Features group of the CommandManager. Click on the Reference Geometry Tool, and choose Plane from the options listed, as shown in Figure 5.5. In the PropertyManager, set the offset distance at 1 inch.

Make sure that the new plane is above the Top Plane, as shown in Figure 5.6 (click on the icon next to the offset distance to change the direction, if necessary), and click the check mark to create the new plane.

FIGURE 5.6

**With the new plane selected (Plane1), select the Line Tool from the Sketch group
of the CommandManager.**

**Draw two horizontal lines, as shown in Figure 5.7, and add an Equal relation to
them. Also add a Vertical relation between two corresponding endpoints.**

FIGURE 5.7

**Select the Tangent Arc Tool. Drag an arc between corresponding endpoints of the
two lines. Be sure to snap to the endpoints of the lines for both arc endpoints, as
shown in Figure 5.8.**

FIGURE 5.8

**Repeat for the other ends of the lines. Select the Centerline Tool, and add the diag-
onal centerline, as shown in Figure 5.9. Add a Midpoint relation between the centerline and the origin. This will center the sketch about the origin. Select the
Smart Dimension Tool, and add the dimensions shown in Figure 5.9 to fully define
the sketch. Close the sketch.**

Both of the sketches required for
the lofted feature are now in place.

FIGURE 5.9

**Click on the Features tab of the
CommandManager. Select the
Lofted Boss/Base Tool, as shown in
Figure 5.10. Now click on each of
the two sketches, near corresponding corners, as shown in Figure
5.11.**

FIGURE 5.10

FIGURE 5.11

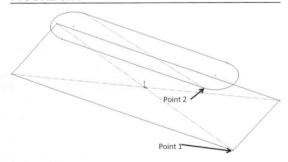

The lofted feature will be created based on a "guide curve." The guide curve is created based on the location of the selection of the sketches, as previewed in Figure 5.12. Click the check mark to complete the loft. From the Heads-Up Toolbar, select Hide/Show Items and View Planes to turn off the display of the construction plane. The resulting part is shown in Figure 5.13.

FIGURE 5.12

FIGURE 5.13

Select the top surface of the part. Select Sketch from the pop-up toolbar, as shown in Figure 5.14, to open a sketch on the selected surface. Select the Offset Entities Tool, as shown in Figure 5.15.

A preview of the offset operation will be displayed, as shown in Figure 5.16.

FIGURE 5.14

FIGURE 5.15

FIGURE 5.16

In the PropertyManager, set the offset distance as 0.20 inches. Check the Reverse box, as shown in Figure 5.17, so that the offset entities are inside of the edges of the face. Click on the check mark to finish.

The finished sketch is shown in **Figure 5.18**. Note that the 0.20-inch offset distance is shown as a dimension that can be edited by double-clicking it.

FIGURE 5.17

FIGURE 5.18

Before cutting the shape of the sketch into the part, we need to consider the design intent. We want this cut to be blind, so that the cards sit on the bottom of the hole, but we want the cut depth to vary with the overall height of the card holder. Therefore, neither a Blind nor Through All type of cut will work. Rather, we will specify the depth of cut so that the bottom of the cut is a fixed distance above the bottom of the card holder.

Select the Extruded Cut Tool from the Features group of the CommandManager. Switch to the Trimetric View. In the PropertyManager, set the type of extrusion to Offset From Surface, as shown in Figure 5.19. As the surface, we want to select the bottom face of the part.

FIGURE 5.19

To select the bottom surface, we could switch to Bottom View or rotate the model until the Bottom View is visible, but in the steps that follow, we will learn a handy technique for selecting a nonvisible surface *without* rotating the model from the Trimetric View.

FIGURE 5.20

Move the mouse above the bottom face. Do not click the left button; doing so would select the visible outer surface. Right-click, and choose Select Other from the menu, as shown in Figure 5.20.

With the bottom face highlighted, as shown in Figure 5.21, click the left mouse button.

If there were other possible selections available, they could be selected from the pop-up menu that is shown in Figure 5.21.

In the PropertyManager, set the offset distance as 0.125 inches, as shown in Figure 5.22, and click the check mark to complete the cut.

FIGURE 5.21

FIGURE 5.22

FIGURE 5.23

The resulting geometry is shown in Figure 5.23.

We will now introduce the Shell command. This command is often used with molded plastic parts to create thin-wall geometries. Since the part we are modeling does not need to have any significant strength, making the part solid would be a waste of material. We will make the wall thickness of the part a constant 0.060 inches.

Select the Shell Tool from the Features group of the CommandManager, as shown in Figure 5.24. Select the bottom face of the part as the face to be removed, as shown in Figure 5.25. In the PropertyManager, set the wall thickness to 0.060 inches, as shown in Figure 5.26. Click the check mark to complete the shell operation, the results of which are shown in Figure 5.27.

FIGURE 5.24

FIGURE 5.25

FIGURE 5.26

FIGURE 5.27

FIGURE 5.28

Select the Fillet Tool from the Features group of the CommandManager, as shown in Figure 5.28. In the PropertyManager, set the fillet radius of 0.10 inches. Select the top face, as shown in Figure 5.29, and click the check mark to apply the fillet.

Note that fillets are usually applied to edges, not faces. Selecting a face causes all of the edges of that face to be filleted, as shown in Figure 5.30.

Select the Fillet Tool again. Select one of the edges at the bottom of the cavity as shown in Figure 5.31 (as long as the "Tangent propagation" box is checked, then the fillet will be extended completely around the bottom edge). Set the radius as 0.050 inches, and click the check mark to apply the fillet.

FIGURE 5.29

FIGURE 5.30

FIGURE 5.31

Although our part appears to be finished, there is a problem that may not be evident from examining the part from standard views. We will use a section view to get a better look at the problem areas.

FIGURE 5.32

Select the Front Plane from the FeatureManager. Click on the Section View Tool in the Heads-Up Toolbar, as shown in Figure 5.32. Click the check mark, and the cross-section of the part is displayed, as shown in a Front View in Figure 5.33.

FIGURE 5.33

Zooming in on the filleted edges shows the problem. Because the filleting was performed after the shell operation, the wall is thinner than desired at the upper corners, as shown in **Figure 5.34.** (Similarly, the wall is thicker than desired in the lower corners.)

FIGURE 5.34

To correct this problem, we could fillet the sharp corners to maintain a constant wall thickness. We could also shell the part *after* creating the fillets. This is the easier way to produce the constant wall thickness. It is not necessary to delete the shell and fillet operations and redo them in the proper order; we can simply reorder them in the FeatureManager.

Click on the Shell in the FeatureManager, as shown in Figure 5.35, and hold down the left mouse button.

Drag the cursor until the arrow points below the two fillets, as shown in Figure 5.36.

Release the mouse button, and the features are reordered, as shown in Figure 5.37.

FIGURE 5.35

FIGURE 5.36

FIGURE 5.37

The wall thickness is now constant, as shown in **Figure 5.38.** Note that not all features can be reordered, as some operations will be based on geometries created by prior operations.

FIGURE 5.38

Click on the Section View Tool again.

This will toggle off the display of the section view.

Text can be added to a part as a sketch entity and then extruded into raised or embossed letters. We will add a part number in raised letters to the bottom of the part.

Click on the Sketch tab of the CommandManager. Switch to Bottom View and open a sketch on the flat surface shown in Figure 5.39 by selecting the Sketch Tool.

Select the Text Tool from the CommandManager, as shown in Figure 5.40. In the PropertyManager, type in the text, as shown in Figure 5.41 (it is not necessary to select a curve to align the text along).

FIGURE 5.39

FIGURE 5.40

FIGURE 5.41

By default, the font specified in the Options will be used. To use another font, click on the "Use document font" box to uncheck it, which will allow you to select the Font button and edit the type and size of the font.

Click in the sketch at the approximate location of the text, and then click the check mark in the PropertyManager to close the text box. The text can now be moved by clicking and dragging it around the sketch. Center the text on the face, as shown in Figure 5.42.

FIGURE 5.42

If necessary, dimensions can be added to the marker at the lower left of the text to positively locate the text.

Click on the Features tab of the CommandManager, and select the Extruded Boss/Base Tool.

FIGURE 5.43

Extrude the sketch 0.020 inches out from the part, as shown in Figure 5.43.

This card holder may be made from a translucent plastic. The SolidWorks program allows parts to be displayed in the color desired, and also for optical properties such as transparency and shininess to be set.

Select the part name from the FeatureManager (so that the entire part is selected), and select Edit: Appearance: Appearance from the main menu, as shown in Figure 5.44.

In the PropertyManager, select a gray color, and move the Transparency slider bar toward the right, as shown in Figure 5.45. Click the check mark to apply the desired color and properties.

The translucent part will look more realistic without the edges displayed, so select shaded (without edges) as the display mode, as shown in Figure 5.46.

FIGURE 5.44

FIGURE 5.45

FIGURE 5.46

The translucent part is shown in Figure 5.47.

Save this part for use in future exercises.

Note that more than two sketches can be used to create a lofted feature. Consider the three sketches shown in Figure 5.48.

FIGURE 5.47

FIGURE 5.48

FUTURE STUDY

Industrial Design

In this chapter, we created a part by using the Loft command to smoothly blend two shapes. Engineers in many industries are sometimes reluctant to use this type of construction, since the resulting surfaces are difficult to define mathematically. This makes the geometry difficult to specify on engineering drawings and often impossible to make with traditional manufacturing methods.

Complex geometries have always been an important tool for *industrial designers*. Industrial designers are important members of product development teams, but perform a different task than do design engineers. The Industrial Designers Society of America defines the role of the industrial designer as:

> The industrial designer's unique contribution places emphasis on those aspects of the product or system that relate most directly to human characteristics, needs and interests. This contribution requires specialized understanding of visual, tac-

tile, safety and convenience criteria, with concern for the user. (www.idsa.org)

The image that many associate with industrial designers is that of an artist working on a clay model of an automobile, creating the shapes that would eventually be seen on the showroom floor. The clay model would eventually be digitized by measuring thousands of points on the surface in order to define the shape for the tooling used to stamp the sheet metal body parts. Industrial designers now create many of their models with the "virtual clay" of solid modeling software, computer-controlled machining centers, and rapid prototyping machines.

The roles of design engineer and industrial designer have begun to overlap as they have access to a similar set of product development tools. However, engineers should recognize and utilize the unique capabilities of industrial designers within the product design process.

The lofted solid created from these sketches is shown in **Figure 5.49**.

Also, an additional sketch defining a *guide curve* can be introduced, allowing more control over the loft. **Figure 5.50** shows a guide curve added to the three previous sketches.

The resulting solid is shown in **Figure 5.51**.

FIGURE 5.49	**FIGURE 5.50**	**FIGURE 5.51**

5.2   Parts Created with Swept Geometry

In this section, we will learn how to create a solid by "sweeping" a cross-section along a path. We will start with a simple part in which the path is planar. Later, we will use a helix curve to form a helical spring. In the next section, we will introduce the 3-D sketch, which will be used to define a sweep path in 3-D space.

The first part that we will create is the bent tubing section shown in **Figure 5.52**. We begin by creating the sweep path. The geometry of the sweep path, which defines the centerline of the tubing, is shown in **Figure 5.53**.

FIGURE 5.52

FIGURE 5.53

Open a new part. Select the Top Plane from the FeatureManager. Choose the Line Tool from the Sketch group of the CommandManager, and draw a vertical line beginning at the origin. Choose the Tangent Arc Tool, and drag out an arc from the endpoint on the line, as shown in Figure 5.54.

Choose the Line Tool. Drag out a line from the endpoint of the arc along the path that is tangent to the arc, as shown in Figure 5.55.

FIGURE 5.54

FIGURE 5.55

A = 56° R = 1.006

1.383

Tangent

FIGURE 5.56

FIGURE 5.57

Select the Smart Dimension Tool, and add the dimensions shown in Figure 5.56. Recall that to create an angular dimension, you select the two lines that define the angle. The sketch should be fully defined. Close the sketch by clicking on the icon in the upper right corner of the graphics area, as shown in Figure 5.57.

Switch to the Front View. Select the Front Plane from the FeatureManager. Select the Circle Tool, and drag out two circles, both centered at the origin. Select the Smart Dimension Tool, and add diameter dimensions of 0.40 and 0.50 inches, as shown in Figure 5.58. Close the sketch.

FIGURE 5.58

FIGURE 5.59

Click on the Features tab of the CommandManager. Select the Swept Boss/Base Tool, as shown in Figure 5.59. In the PropertyManager, select the sketch containing the two circles (Sketch2) as the Profile, the cross section to be swept, and select the first sketch defining the geometry of the tubing centerline (Sketch1) as the Path of the sweep, as shown in Figure 5.60. A preview of the swept geometry is shown in Figure 5.61. Click the check mark to complete the operation.

FIGURE 5.60

Profile
Path

FIGURE 5.61

The completed part is shown in **Figure 5.62**.

FIGURE 5.62

In the next exercise, we will use a more complex sweep path, a helix, to create a helical spring.

Open a new part. Select the Front Plane from the FeatureManager. Select the Circle Tool from the Sketch group of the CommandManager. Draw a circle centered at the origin. Select the Smart Dimension Tool, and dimension the circle's diameter as 2 inches, as shown in Figure 5.63.

FIGURE 5.63

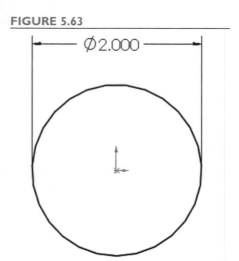

From the main menu, select Insert: Curve: Helix/Spiral, as shown in Figure 5.64.

A helix can be defined by specifying any two of the following three quantities:

1. The height, or overall length of the helix,

2. The pitch, the distance between similar points on successive turns of the helix, and

3. Revolutions, the total number of complete turns of the helix.

FIGURE 5.64

We will define the height and the number of revolutions and allow the pitch to be calculated.

FIGURE 5.65

Select Height and Revolution in the "Defined By" box. Set the height to 6 inches, the number of revolutions to 8, and the starting angle to 135 degrees, as shown in Figure 5.65.

FIGURE 5.66

A preview of the helix geometry is shown in **Figure 5.66**. The start angle is not critical here. If the start angle is set to a multiple of 90 degrees, then we can define our profile sketch in either the Right or the Top Plane. We have chosen an arbitrary angle to illustrate the procedure for creating a plane at the end of a path sketch.

Click the check mark to accept the helix definition and close the sketch.

We will now create a new plane at the end of the helix, perpendicular to the helix at that point.

FIGURE 5.67

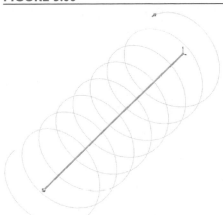

Click on the Features tab of the CommandManager. Click on the Reference Geometry Tool, and choose Plane, as shown in Figure 5.67. Click once on the helix curve, and then click on the endpoint of the curve, as shown in Figure 5.68. In the PropertyManager, note that "Normal to Curve" has been automatically selected as the plane definition, as shown in Figure 5.69. Click the check mark to create the plane.

FIGURE 5.68

FIGURE 5.69

Click the Sketch tab of the CommandManager. With the new plane selected, choose the Circle Tool. Drag out a circle, as shown in Figure 5.70. Click the check mark to complete the circle.

It is not possible to snap the center of the circle to the endpoint of the helix. Rather, we will use a *pierce* relation. The pierce relation is defined between a point and a curve, and sets the point at the location where the curve "pierces" the sketch containing the point.

FIGURE 5.70

FIGURE 5.71

Click on the center point of the circle to select it. While holding down the Ctrl key, select the helix near the endpoint, as shown in Figure 5.71. In the PropertyManager, click on Pierce, as shown in Figure 5.72. Click the check mark to add the relation. Choose the Smart Dimension Tool, and add a 0.25-inch diameter dimension to the circle. The sketch should now be fully defined, as shown in Figure 5.73. Close the sketch.

FIGURE 5.72

FIGURE 5.73

Click on the Features tab of the CommandManager. Choose the Swept Boss/Base Tool. In the PropertyManager, choose the sketch containing the 0.25-inch-diameter circle as the Profile, and the helix as the Path, as shown in Figure 5.74. Click the check mark to complete the sweep operation.

The completed spring is shown in **Figure 5.75**. In this exercise, we used a curve in 3-D space as the sweep path. This curve was created from a 2-D sketch (a circle). In the next section, we will use the more general 3-D Sketch Tool to define the sweep path in three-dimensional space.

FIGURE 5.74

FIGURE 5.75

5.3 A Part Created with a 3-D Sketch as the Sweep Path

The use of a 3-D sketch as a sweep path allows more complex parts to be created. As the name implies, a 3-D sketch contains entities in 3-D space, whereas typical sketches contain entities that exist in a plane. Not all sketch entities are available in a 3-D sketch.

In this exercise, we will model the handlebars shown in **Figure 5.76**. Since the handlebars are defined in metric units, we will need to set the units accordingly.

Open a new part. Select Tools: Options from the main menu. Under the Document Properties tab, select Units, and choose MMGS (millimeters, grams, seconds). Set the number of decimal places for length units to None.

From the Sketch group of the CommandManager, click the arrow under the Sketch Tool and select 3D Sketch, as shown in Figure 5.77.

FIGURE 5.76

FIGURE 5.77

When working with 3-D sketches, the axes displayed in the corner of the screen, shown in **Figure 5.78**, are especially important. When we draw lines, we will do so in one of the primary planes—XY, YZ, or ZX.

Select the Line Tool.

Note that beside the cursor, the plane that the line will be drawn in is displayed, as shown in **Figure 5.79**. We want our first two lines to be sketched in the ZX plane, so we will change this orientation before proceeding.

Press the Tab key, which causes the sketch plane to cycle between the three principal planes. Stop when the plane selected is ZX, as shown in Figure 5.80.

FIGURE 5.78   FIGURE 5.79   FIGURE 5.80

Drag out a line from the origin along the X axis, as shown in Figure 5.81**. Make sure that the "Along x-axis" relation icon is displayed before releasing the mouse button.**

From the endpoint of the first line, drag out a diagonal line, as shown in Figure 5.82**.**

When dimensioning a 3-D sketch, not all of the options available with 2-D sketches can be used. For example, if we try to add a dimension from the origin to the endpoint of the second line, only the straight-line distance between the two points can be displayed. If we want to dimension the x and y distances from the origin to the point, then the use of a centerline is necessary.

FIGURE 5.81

FIGURE 5.82

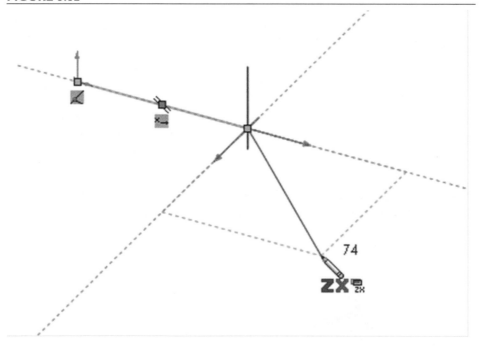

Select the Centerline Tool from the Line Tool pull-down menu. Make sure that the ZX designation still shows beside the cursor. Drag out a centerline from the origin along the Z axis, as shown in Figure 5.83.

FIGURE 5.83

Along z-axis

FIGURE 5.84

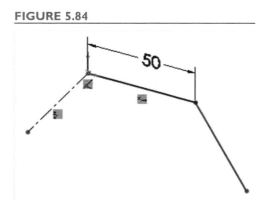

Select the Smart Dimension Tool. Click once on the first solid line drawn, and place a 50 mm dimension, as shown in Figure 5.84.

Add other dimensions between the last endpoint of the second solid line and the centerline, and between the last endpoint of the second solid line and the first solid line, as shown in Figure 5.85.

FIGURE 5.85

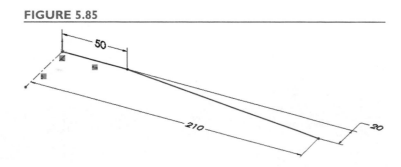

The other lines of the sketch will be drawn in the YZ plane.

FIGURE 5.86

Select the Line Tool. Press the Tab key until YZ is shown as the drawing plane, as shown in Figure 5.86. (If the plane does not change when you press the Tab key, then click once on the axis display in the corner to "reset" this feature.)

Drag out a line from the last endpoint of the second solid line in the Z-direction, as shown in Figure 5.87.

Drag a line downward (in the –Y direction), as shown in Figure 5.88.

FIGURE 5.87

FIGURE 5.88

Drag the last line in the –Z direction, as shown in Figure 5.89.

Select the Smart Dimension Tool. Add the dimensions shown in Figure 5.90.

Note that the sketch is not fully defined. 3-D sketches are more difficult to fully define than are 2-D sketches. Although a line may be drawn in a specific plane, it is not *locked* into that plane unless it is aligned with one of the principal axes. We

FIGURE 5.89

FIGURE 5.90

could fix some of the endpoints to fully define the sketch, but that is not necessary.

We will now add fillets to the sharp corners of our sketch. The fillet radii to be added are shown in **Figure 5.91**.

Select the Sketch Fillet Tool, as shown in Figure 5.92. Add the fillets one at a time. The finished sketch is shown in Figure 5.93.

Close the sketch.

FIGURE 5.91

FIGURE 5.92

FIGURE 5.93

Select the Right Plane from the FeatureManager. Select the Circle Tool. Drag out two circles from the origin, as shown in Figure 5.94.

Add diameter dimensions of 25.4 mm and 23.4 mm, as shown in Figure 5.95. To display the dimensions to one decimal place, select each dimension and choose 0.1 as the precision in the PropertyManager.

FIGURE 5.94

FIGURE 5.95

(Although most bicycle components are specified in metric units, a 1-inch outer handlebar diameter, 25.4 mm, is a standard size.)

Close the sketch.

Click on the Features tab of the CommandManager. Select the Swept Boss/Base Tool. In the Property-Manager, select the 2-D sketch just completed as the profile and the 3-D sketch as the path, as shown in Figure 5.96.

FIGURE 5.96

Click the check mark to complete the sweep. The result is shown in Figure 5.97.

FIGURE 5.97

Select the Mirror Tool, as shown in Figure 5.98. In the Property-Manager, select the Right Plane as the mirror plane, as shown in Figure 5.99, and click the check mark.

FIGURE 5.98

FIGURE 5.99

FIGURE 5.100

The completed handlebars are shown in **Figure 5.100**.

We will now determine the mass of the handlebars, which will be made from 6061 aluminum.

Right-click on Material in the FeatureManager. Select Edit Material and select 6061 Alloy from the list of aluminum alloys. Click Apply and Close.

Select Tools: Mass Properties from the Main Menu, as shown in Figure 5.101.

FIGURE 5.101

The mass of the handlebars is calculated as about 221 grams, as shown in Figure 5.102.

FIGURE 5.102

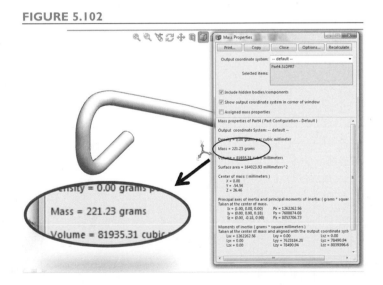

PROBLEMS

P5.1 Create the solid object shown in **Figure P5.1** with a loft between a square base and a circular top (dimensions are in inches).

FIGURE P5.1

P5.2 Create the part shown in **Figure P5.2A**. Create the sketch shown in **Figure P5.2B** in the Top Plane. Create a new plane 4 inches above the Top Plane, and create the second sketch consisting of the two arcs and two lines indicated in **Figure P5.2C**, snapping to corresponding points of the first sketch. Create a loft between the two sketches to finish the part. All dimensions are in inches.

FIGURE P5.2A

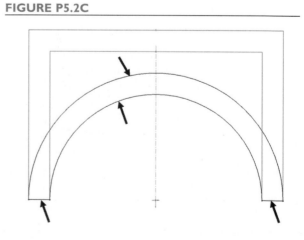

FIGURE P5.2B

FIGURE P5.2C

P5.3   Create the part shown here, with a circular cross-section of
1-inch diameter. All dimensions are in inches.

FIGURE P5.3A   **FIGURE P5.3B**

R4.00

10.00

P5.4   Create the shape shown in **Figure P5.4A** as a loft defined by the two
ellipses and one circle shown in **Figure P5.4B**. All dimensions are in
inches.

FIGURE P5.4A   **FIGURE P5.4B**

⌀1.00

2.00

Third Loft Guide Point

1.50

3.00

Second Loft Guide Point

2.20

1.00

First Loft Guide Point

2.00

To create an ellipse, select the Ellipse Tool from the Sketch group of the CommandManager. Drag out a circle from the origin, as shown in **Figure P5.4C**, and then click and drag a point on the edge of the circle to "flatten" it into an ellipse, as shown in **Figure P5.4D**. Snap points are created at four locations on the ellipse; use these points to define the semimajor and semiminor axes of the ellipse. Add centerlines from the origin to the snap points, as shown in **Figure P5.4E**, and set them to horizontal and vertical to correctly align the ellipse and fully define the sketch.

When defining the 1-inch circle for the top of the loft, add a center line to the horizontal quadrant point, as shown in **Figure P5.4F**. This will provide the final guide point for the loft (see **Figure P5.4B**).

FIGURE P5.4C

FIGURE P5.4D

FIGURE P5.4E

FIGURE P5.4F

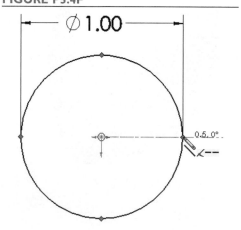

P5.5 Turn the shape created in **Problem P5.4** into the bottle shown in **Figure P5.5A**. Extrude a circular neck, as shown in **Figure P5.5B**, add a 0.25-inch fillet to the neck-to-body junction, and use the Shell Tool to hollow the bottle, leaving a 0.020-inch wall thickness. What is the volume of the space within the bottle? (Hint: To find the volume, move the rollback bar to just before the shell command, as shown in **Figure P5.5C** and find the volume of the solid before shelling. Then move the rollback bar past the shell, and find the volume of the bottle itself. The difference in the two volumes is the volume contained within the bottle.)

FIGURE P5.5A

FIGURE P5.5B

FIGURE P5.5C

P5.6 Model a bent tube with 0.5-inch outer diameter and 0.375-inch inner diameter to join the ends of the two tubes shown in **Figure P5.6A**. Follow the route shown by the centerlines in **Figure P5.6A**, and add 1-inch radius fillets to the corners.

The completed model is shown in **Figure P5.6B**.

FIGURE P5.6A

FIGURE P5.6B

P5.7   Repeat Problem P5.6, using a different path. Use a tube geometry that minimizes the tube length while maintaining a straight section at each end, as shown in **Figure P5.7**.

FIGURE P5.7

6.000

4.000

5.000

P5.8   In Chapter 4, we noted that screw threads are not usually modeled. Instead, cosmetic threads are added to represent the threads. In some cases, however, a threaded part needs to be prototyped, so the threads must be modeled. **Figure P5.8A** shows the drive screw from a table vise. The threads on a drive screw are typically Acme threads, a trapezoidal-profile thread. The thread shape can be seen in the cross-section of the screw of **Figure 5.8B**. Model the screw, following these steps:

FIGURE P5.8A

FIGURE P5.8B

1. Create a cylinder 0.5 inches in diameter and 6 inches in length.
2. Create a new plane 0.1 inches offset from one end of the cylinder.
3. Sketch a 0.5-inch diameter circle in the new plane, and insert a clockwise helix with a pitch of 0.1 inches (10 threads per inch) and a height of 6.2 inches, as shown in **Figure 5.8C**.

FIGURE P5.8C

4. Open a new sketch in a plane normal to the helix at its endpoint. Sketch the profile as shown in **Figure P5.8D**, and close the sketch.

5. Select the Swept Cut Tool. Sweep the profile along the helix to cut the threads.

6. Add the head of the screw, using the dimensions shown in **Figure P5.8E**.

FIGURE P5.8D

FIGURE P5.8E

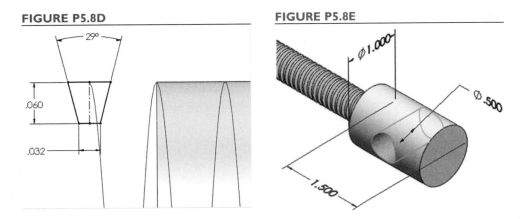

P5.9 Measure a standard paper clip, and use your measurements to create a solid model similar to the one shown in **Figure P5.9**.

FIGURE P5.9

CHAPTER 6

Building Assembly Models from Part Models

Introduction

In the preceding chapters, the development of solid models of parts was covered in detail. In this chapter, methods for combining such part models into complex, interconnected solid models will be described. These types of models, composed of interconnected part models, are called *assembly models*.

The assembly that will be constructed in this chapter is a model of a hinged hatch, as shown in **Figure 6.1**.

The chapter will begin with a tutorial describing the construction of the part models of four components used in this assembly. After the models are constructed, they will be interconnected first into a simple subassembly, and then into a more complex assembly model. The subassembly and assembly models will be used in subsequent chapters to demonstrate some advanced assembly modeling features.

FIGURE 6.1

Chapter Objectives

In this chapter, you will:

- create holes using the Hole Wizard,

- learn to import part models into an assembly,

- define assembly mates between parts, and

- create subassemblies, and use them within larger assemblies.

6.1 Creating the Part Models

Before an assembly can be created, the parts to be assembled must be modeled. In this first step, solid models of the four components will be created.

The first model that we will create is the hinge component.

Start a new part model, and sketch a horizontal line from the origin and a tangent arc in the Front Plane, as shown in Figure 6.2.

Add the dimensions shown in Figure 6.3. Add a centerline from the center of the arc to the endpoint of the arc. In the PropertyManager add a horizontal constraint to the centerline, as shown in Figure 6.3.

FIGURE 6.2

FIGURE 6.3

FIGURE 6.4

The sketch should now be fully defined. While in previous exercises we have used sketches with closed contours to make extrusions, in this case we will use the open-contour sketch to create a *thin-feature* extrusion.

Create an Extruded Base. Note that the Thin Feature box is checked, since the sketch contour is open. Set the extrusion depth at 4 inches and the thickness at 0.25 inches, as shown in Figure 6.4. Change the directions as necessary so that the extrusion and thickness directions are as shown in the preview in Figure 6.5.

The completed extrusion is shown in Figure 6.6.

FIGURE 6.5

FIGURE 6.6

FIGURE 6.7

Select the top horizontal surface of the hinge, as shown in Figure 6.7, for the next sketch. Choose the Normal To View.

Sketch the 1-inch square area to be cut away from the basic hinge shape, as shown in Figure 6.8. Use snaps and/or relations to align the edges of the square with the edges of the hinge, so that the sketch is fully defined with only the single dimension shown.

FIGURE 6.8

Use the Extruded Cut Tool to cut Through All in both Direction 1 and Direction 2, as shown in Figure 6.9.

The result of the cut is shown in Figure 6.10.

FIGURE 6.9

Select the extruded cut from the FeatureManager, and define a linear pattern to create another instance of this feature 2 inches along the horizontal edge of the hinge, as shown in the preview in Figure 6.11.

FIGURE 6.10

The completed pattern is shown in Figure 6.12. A pattern of countersunk screw holes will now be added to the hinge. Since fastener holes are generally of standard dimensions, an intelligent design tool known as the Hole Wizard will be used to create the holes.

FIGURE 6.11

FIGURE 6.12

FIGURE 6.13

Choose the Hole Wizard Tool from the Features group of the CommandManager, as shown in Figure 6.13.

The Hole Wizard dialog box appears. The Hole Wizard can be used to create holes to accommodate most standard fastener types. We will create countersunk holes for #10 flat head wood screws. (Screw diameters are designated by a number from 1 to 12, after which they are designated by the diameter as a fraction of an inch. A #10 screw has a diameter of 0.190 inches.)

FIGURE 6.14

In the Hole Specification PropertyManager under the Type tab, click to set the hole specification to Countersunk. Set the Standard to Ansi Inch, the Type to Flat Head Screw (82), the size to #10, and the End Condition to Through All, as shown in Figure 6.14. Click the Positions tab in the PropertyManager to initiate the Hole Position PropertyManager.

We are now prompted to enter the hole location, as shown in Figure 6.15. We will create a single hole, and replicate it using a linear pattern.

FIGURE 6.15

FIGURE 6.16

Change to the Top View. Place the center of the hole by clicking in the approximate location shown in Figure 6.16.

Select the Smart Dimension Tool, and add dimensions to the hole location as shown in Figure 6.17.

Click the check mark twice to create the hole, which is shown in Figure 6.18.

FIGURE 6.17

FIGURE 6.18

Select the new hole from the FeatureManager, and create a linear pattern with four total instances of the hole, spaced 2 inches along the long side of the hinge and 0.9 inches along the short side, as shown in the preview in Figure 6.19.

The final model of the hinge is shown in **Figure 6.20**. Since it will be used in a later assembly it must be saved.

FIGURE 6.19

FIGURE 6.20

Save this using the file name "Hinge," and close the file.

Now, the second major component in the first subassembly will be created.

Open a new part, and sketch a 16-inch by 16-inch square in the Front Plane, centered about the origin. Extrude it 2.5 inches.

This base feature is shown in **Figure 6.21**.

On the front face, sketch a 10-inch square centered about the origin. Extrude it 1 inch from the face.

This completes the second component, as shown in **Figure 6.22**. When we assemble the parts later, it will be helpful if they are different colors.

FIGURE 6.21

FIGURE 6.22

FIGURE 6.23

If desired, right-click on the part name from the FeatureManager, and select Appearance: Appearance, as shown in Figure 6.23. Pick the new color for the hatch, and click the check mark to close the Color PropertyManager.

Save it using the file name "Hatch." Close the file.

A model of the hinge pin will now be created.

Open a new part, and sketch a 0.5-inch diameter circle in the Front Plane, centered at the origin. Extrude it 4 inches. Add a 1-inch diameter cap on the pin, extruding it 0.25 inches.

FIGURE 6.24

The pin is shown in Figure 6.24. Again, using a new color for this component will aid in visualization of the final assembly.

Change the color of the part, if desired, and save the part in a file named "Pin." Close the file.

The final component needed for the assembly is a frame.

Open a new part, and in the Front Plane sketch a 25-inch square centered at the origin. Extrude it to a depth of 4 inches. Extrude a 14-inch square cut (centered on the front face of the component), yielding the part shown in Figure 6.25.

FIGURE 6.25

Change the color, if desired. Save this part in a file named "Frame," and close it.

These four parts will now be used to create a model of the hatch assembly.

6.2   Creating a Simple Assembly of Parts

The features of the software that we will employ are the *assembly* capabilities. *Assemblies* are complex solid models that are made up of simpler part models, with specifically defined geometric relationships between the parts. The SolidWorks program provides us with the ability to relate surfaces and other geometric features of one part to those of another part as follows:

- *Define two flat surfaces as **coincident**:* This places the two flat surfaces in the same plane.
- *Define two flat surfaces as **parallel**.*
- *Define two flat surfaces a preset **distance** apart:* This makes the two surfaces parallel, with a specified distance between them.

- *Define two lines or planes as **perpendicular** to one another.*
- *Define two lines or planes at a preset **angle** to one another.*
- *Define a cylindrical feature as **concentric** with another cylindrical feature:* This aligns the axes of two cylindrical features.
- *Define a cylindrical feature as **tangent** to a line or plane.*

These geometric relationships are called *mates*. There are many other geometric relationships that can be accommodated as well.

In this section, a tutorial will be presented in which we will create a simple assembly by attaching a set of hinges to the hatch component, as shown in **Figure 6.26**. This assembly will be used in the following section as a small assembly (subassembly) within a larger assembly.

FIGURE 6.26

FIGURE 6.27

To begin creating an assembly model, select File: New from the main menu. Click on the Assembly icon (Figure 6.27), and click OK.

A new assembly window will be created. For the purposes of this assembly, the main "base" part will be the hatch. We will begin by importing this component into the assembly.

FIGURE 6.28

In the PropertyManager, any open parts will be displayed. Since we have closed our part files, click on Browse to find the hatch file, as shown in Figure 6.28. (Note: If the box shown does not appear when you start a new assembly, select Insert: Component: Existing Part/Assembly from the main menu.)

Browse to the location where the hatch file was saved, as shown in Figure 6.29, and open it. If necessary, change the file type option to Part (*.prt, *.sldprt) to find the part files.

FIGURE 6.29

DESIGN INTENT | Planning an Assembly Model

When we imported the first part into our assembly, we were careful to locate the part at the origin of the assembly space. More precisely, we made the origin of our part model coincident with the origin of the assembly space, and therefore also made the Front, Top, and Right planes of our part model coincident with the corresponding planes in the assembly space. This is not strictly necessary; we could have located the origin of the part model anywhere in the assembly space. However, by taking advantage of the default Front, Top, and Right planes (as well as the origin), we can use these as references for the addition of new features (such as holes, bosses, etc.) at the assembly level (as we will do in Chapter 7), as well as in construction of assembly drawings (as we will do in Chapter 8). Judicious choice of the location of the first part we bring into an assembly can simplify subsequent tasks, if we anticipate our future use of the assembly model.

FIGURE 6.30

As prompted by the message shown in Figure 6.30, click the check mark (the OK button) to insert the hatch part at the origin.

This will import the hatch model into the assembly, with the origin of the hatch coincident with the origin of the assembly window.

From the Heads-Up View toolbar, click the Apply Scene Tool and select Plain White as the background, as shown in Figure 6.31.

Note that the name of the component (Hatch) now appears in the FeatureManager (Figure 6.32). The designation (f) means that the component is "fixed"; it is fully constrained in the assembly window, and cannot be moved or rotated. Note the Mates group at the bottom of the FeatureManager. As we define the geometric relations between components, these relations will be stored under this Mates group.

The hinge component will now be brought into the assembly. Note that the Command-Manager has an Assembly group containing many of the tools we will use in assembling the components.

FIGURE 6.31

FIGURE 6.32

FIGURE 6.33

FIGURE 6.34

Select the Insert Component Tool from the Assembly group of the CommandManager, as shown in Figure 6.33. Select Browse, and find the hinge file. Click Open, and move the hinge to the approximate position shown in Figure 6.34. Click to place the hinge.

This will import the hinge component, with its origin located at the point selected. The exact position and orientation are not important at this point, since the Mate Tool will be used to establish the position and orientation with respect to the base part.

Note that the name of the hinge component now appears in the FeatureManager, with the (–) designation preceding it. This designation indicates that the component is "floating," and can be moved or rotated (as its degrees of freedom allow).

Although the CommandManager contains Move Component and Rotate Component Tools, parts can also be moved or rotated directly with click-and-drag operations.

Click on the hinge with the left mouse button, and, while holding the button down, drag the hinge to a new position, as shown in Figure 6.35. Hit Esc to exit this mode.

Click on the hinge with the right mouse button and, while holding the button down, rotate the hinge to a new orientation, as shown in Figure 6.36. Hit Esc to exit this mode.

FIGURE 6.35

FIGURE 6.36

More control over the move and rotate commands is available with the Triad Tool. The Triad Tool can be activated from the right-click menu of a component.

Right-click on the hinge and select Move with Triad from the menu, as shown in Figure 6.37.

FIGURE 6.37

The Triad Tool is shown in **Figure 6.38**. The arrows correspond to the principal axes of the assembly. Click and drag on one of the arrows, as shown in **Figure 6.39**, to translate the part in that direction. Click and drag on one of the circles, as shown in **Figure 6.40**, to rotate the part within the plane defined by that circle.

FIGURE 6.38

FIGURE 6.39

FIGURE 6.40

FIGURE 6.41

Experiment with moving and rotating the hinge component with the Triad Tool. Place the hinge in the approximate position and orientation shown in Figure 6.41. Click in the white space around the hinge to turn off the Triad Tool.

While it is not strictly necessary to place the component in its approximate position and orientation prior to defining mate instructions, doing so can remove ambiguity and simplify the establishment of the mates.

The Mate Tool will now be used to establish the first geometric relationship between the hinge and the hatch.

FIGURE 6.42

Click the Mate Tool from the Assembly group of the Command-Manager, as shown in Figure 6.42.

This brings up the Mate PropertyManager.

Using the Rotate View Tool, as shown in Figure 6.43, rotate the view so that the bottom face of the hinge can be seen. Press the Esc key to turn off the Rotate View Tool. Select the bottom face (Figure 6.44).

FIGURE 6.43

FIGURE 6.44

The name of the surface will appear in the highlighted area of the Mate dialog box.

If you select something incorrectly during a mate operation, you can clear the selection box at any time by right-clicking in the assembly window and selecting the Clear Selections option.

FIGURE 6.45

Return to a Trimetric or Isometric View, and select the top face of the hatch component (Figure 6.45).

In the Mate dialog box, the selected faces are shown, and a list of possible mates is shown (**Figure 6.46**). By default, a coincident mate is selected when two flat surfaces are selected. This means that the two selected faces will be coplanar. The hinge will move to satisfy the selected mate configuration, as shown in **Figure 6.47**.

FIGURE 6.46

The Mate Alignment Tools in the dialog box are important because there is often more than one configuration that meets the specification of the selected mate. For example, the hinge could be upside-down and the selected mate could still be satisfied. An advantage of placing and orienting a com-

FIGURE 6.47

ponent before applying mates is that the default alignments of the mates are usually correct. However, we will illustrate the use of the Mate Alignment Tools before applying the mate to the hatch and hinge.

FIGURE 6.48

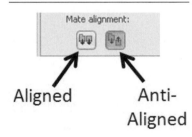

Toggle between the Aligned and Anti-Aligned Tools, as shown in Figure 6.48. Note that the hinge is flipped, as shown in Figure 6.49. Choose the tool which results in the proper alignment.

To better view the effect of the mate, switch to the Right View.

As shown in **Figure 6.50**, the bottom of the hinge and the top of the hatch are coplanar.

FIGURE 6.49

FIGURE 6.50

FIGURE 6.51

Click the check mark in the Mate dialog box or the pop-up box to apply the mate.

Note that the hinge component can be moved with respect to the fixed hatch component, but that the two mated faces remain coplanar. That is the geometric effect of the mate.

A second mate will now be added to provide additional location information for the hinge relative to the fixed hatch component.

Rotate the view orientation so that the back face of the hatch is visible. Click and drag the hinge to the approximate position shown in Figure 6.51.

With the Mate dialog box still open (select the Mate Tool if you closed it accidentally), select the two faces shown in Figure 6.52. Click the check mark to apply this second coincident mate.

The result of the mate is shown in **Figure 6.53**.

FIGURE 6.52

FIGURE 6.53

Select the two faces shown in Figure 6.54.

In this case, we do not want to apply the default coincident mate. Rather, we want these faces to be a specific distance apart.

FIGURE 6.54

Select a Distance mate from the Mate dialog box or the pop-up box, as shown in Figure 6.55. Set the distance to 1 inch, as shown in Figure 6.56.

FIGURE 6.55

FIGURE 6.56

A distance mate has two possible configurations. If you check the Flip Dimension box, then the other configuration is selected, as shown in **Figure 6.57**.

With the mate defined correctly, as in Figure 6.58, click the check mark to apply the mate.

FIGURE 6.57

FIGURE 6.58

Click Esc to close the Mate dialog box. Try to move the hinge by clicking and dragging it.

A message appears, as shown in **Figure 6.59**, stating that the part cannot be moved. The three mates that we have applied have completely defined its position and orientation relative to the fixed hatch.

A second hinge will now be added to the assembly. Since the second hinge will be identical to the first, no new component needs to be created. A second instance of the first hinge can simply be added to the assembly.

FIGURE 6.59

The selected component is fully defined. It cannot be moved.

FIGURE 6.60

Select the Insert Component Tool, and browse to the hinge file. Place the hinge into the assembly window, as shown in Figure 6.60.

Using a procedure similar to the one outlined previously in this chapter, add two coincident mates and a distance mate to fully constrain the second hinge in the position shown in Figure 6.61.

FIGURE 6.61

Once defined, mates can be easily modified. For instance, assume that a redesign specified that the location of the second hinge should be 4 inches from the mated edge rather than 1 inch.

FIGURE 6.62

Click the + sign next to the Mates entry in the FeatureManager. Locate the distance mate associated with the second hinge (called Distance2), and right-click on the mate name, as shown in Figure 6.62. Select the Edit Feature option.

The PropertyManager associated with this mate will reopen, allowing for editing of the mate.

FIGURE 6.63

Change the value of the distance to 4.00 inches, as shown in Figure 6.63. Accept the change by clicking the check mark.

The resulting position of the second hinge is shown in Figure 6.64.

FIGURE 6.64

In this way, a parameter of the mate can be readily redefined. Since we will want to continue with the 1-inch value rather than the modified 4-inch value, we will revert to the previous value.

Press the Esc key to close the Mate dialog box. Click the Undo Tool, shown in Figure 6.65, to revert to the 1-inch dimension. Save the assembly file using the name "Door," and close the file.

FIGURE 6.65

Note that the new file has the extension .SLDASM, indicating that it is a SolidWorks assembly file.

6.3 Creating a Complex Assembly of Subassemblies and Parts

FIGURE 6.66

In this section, we will create a new assembly using the door assembly created in the previous section as a subassembly. The assembly will be a model of a hinged hatch, shown in Figure 6.66.

The assembly will also make use of the hinge, pin and frame parts created in Section 6.1.

FIGURE 6.67

Open a new assembly. If the Insert Component PropertyManager does not open automatically, select the Insert Component Tool from the Assembly group of the CommandManager. Browse to the frame file, and place it at the origin (Figure 6.67).

In the next step, the door assembly created in the previous section will be used as a subassembly within the new assembly. An assembly can be inserted just like a part.

Select the Insert Component Tool, and click the Browse button. In the dialog box, change the file type option to Assembly files, or All Files, and locate the file Door.SLDASM. Select it, and click Open (Figure 6.68).

FIGURE 6.68

FIGURE 6.69

Drop the subassembly into the assembly window in the approximate position shown in Figure 6.69 by clicking.

While the door is an assembly and not a part, it can be handled (mated, rotated, moved, etc.) just like a part component in the assembly window.

Right-click and drag on the door subassembly to rotate it into the approximate orientation shown in Figure 6.70, with the curved portion of the hinges toward the frame.

In our previous mates, we have established relationships between surfaces of the components to be mated. While this is often the procedure we will use, it is sometimes preferable to use the default (Front, Top, or Right) planes, the origins, or other reference geometries in our mates. In centering the Door subassembly with respect to the frame, we will establish a coincident mate between the Right Plane used in the Door sub-assembly model and the Right Plane used in the Frame part model.

Select the Mate Tool. Expand the FeatureManager "flyout" by clicking the plus sign, and expand the entries for Hatch and Door. Select the Right Plane from each, and establish a coincident relation between them, as shown in Figure 6.71. Click the check mark twice to apply the mate and close the Mate Tool.

FIGURE 6.70

FIGURE 6.71

Instances of the hinge parts will now be added to the frame.

Select the Insert Component Tool. Choose Browse, and change the file type option back to part files. Select the hinge part. Drop it into the assembly, and rotate and move the hinge to place it into the approximate location and orientation shown in Figure 6.72.

Click on the Mate Tool to begin a mate. Instead of rotating the view orientation to select the bottom face of the hinge, move the cursor over the hinge, right-click, and choose Select Other, as shown in Figure 6.73. With the back face highlighted, as shown in Figure 6.74, click the left mouse button.

FIGURE 6.72

FIGURE 6.73

FIGURE 6.74

Select the front face of the frame (Figure 6.75). Click the check mark to apply a coincident mate.

The next mate will align the faces of the two hinges to be engaged.

Select the face shown in Figure 6.76 on the hinge that was just mated to the frame.

FIGURE 6.75

FIGURE 6.76

FIGURE 6.77

Also select the corresponding face on the hinge mounted to the hatch, as shown in Figure 6.77. Click the check mark to apply a coincident mate.

The effect of the mate can be most easily seen from the Front View, as shown in **Figure 6.78**.

These two faces now lie in the same plane, but additional mates are required to fully constrain the desired relationship between these components. The hinges will now be brought together to share a common axis of rotation.

FIGURE 6.78

Select the cylindrical faces of each of the two hinges to be mated (Figure 6.79).

By default, a concentric mate will be previewed between the two cylindrical faces, as shown in **Figure 6.80**.

Click the check mark to apply the mate.

FIGURE 6.79

FIGURE 6.80

The final mate required will locate the hinge on the frame.

If necessary, click and drag the hatch downward until the top of the hinge is below the top of the frame.

Select the top surfaces of the hinge and frame, and apply a 2-inch distance mate, as shown in Figure 6.81. Close the Mate PropertyManager.

The hinge mate is now fully defined. We can see the effect of the mates by attempting to move the hatch.

Click and drag on the hatch, as shown in Figure 6.82.

Note that the hatch has a degree of freedom about the hinge, but is constrained against all other types of motion.

FIGURE 6.81

FIGURE 6.82

FIGURE 6.83

Note also that the hatch can apparently rotate through the frame, as shown in Figure 6.83. This is due to the fact that only geometric relations between the entities have been defined, but no true physical characteristics have been imparted to the objects. (We will learn how to use collision detection to limit the motion of the model in Chapter 7.)

Insert a second hinge part, and add mates to place it as shown in Figure 6.84.

We will now add pins to the assembly.

Choose the Insert Component Tool. Click on the pushpin icon, in Figure 6.85. This will enable the insertion of multiple parts. Browse to find the pin, and click two loca-

FIGURE 6.84

tions to place pins in the assembly, as shown in Figure 6.86. **Click the check mark to close the Insert Component PropertyManager.**

FIGURE 6.85

To aid in the selection of small details, a *filter* is sometimes used. When a filter is active, only certain entities can be selected. We are adding mates between faces, so a filter that allows only faces to be selected will be helpful.

FIGURE 6.86

Right-click on the menu bar or CommandManager. From the list of available toolbars, select Selection Filter, as shown in Figure 6.87.

In the Selection Filter toolbar, select the Filter Faces Tool, as shown in Figure 6.88.

Whenever a filter is active, a filter icon appears beside the cursor, as shown in Figure 6.89.

Select the Mate Tool, and select the cylindrical face of the pin, as shown in Figure 6.90. (Note that with the filter set, you cannot select an edge.)

FIGURE 6.87

FIGURE 6.88

FIGURE 6.89

FIGURE 6.90

Select one of the cylindrical faces of the hinge, as shown in Figure 6.91. (Either the outer or the inner face may be selected; since they are concentric, the resulting mate will be the same.)

The pin and the hinge will align concentrically in the preview of the mate. If necessary, change the alignment (from anti-aligned to aligned, or vice versa) of the mate to orient the head of the pin appropriately, and drag the pin to the position shown in Figure 6.92. Click the check mark to apply the mate.

Select the face on the hinge shown in Figure 6.93, and then the underside of the head of the pin, as shown in Figure 6.94.

Click on the check mark to apply the mate, the result of which is shown in Figure 6.95. Close the Mate PropertyManager.

Using this procedure, duplicate the assembly of the second pin.

The assembly is now complete, as shown in Figure 6.96. It will be used in subsequent chapters, so it must be saved. However, it is a good idea to clear any filters when they are no longer needed.

FIGURE 6.91

FIGURE 6.92

FIGURE 6.93

FIGURE 6.94

FIGURE 6.95

FIGURE 6.96

Select the Clear All Filters Tool, as shown in Figure 6.97. Click on the X in the corner of the toolbar to close it.

The Filter Faces Tool is very useful when working with mates. Because it is used so often, there is a keyboard shortcut that can be used to activate it. Pressing the X key will activate the Filter Faces Tool; pressing the F6 key will deactivate the tool. The E and V keys can be used to activate the Filter Edges Tool and Filter Vertices Tool in a similar manner.

FIGURE 6.97

Save the assembly in a file entitled "Hatch Assembly."

PROBLEMS

P6.1   Create a 6.5 inch x 2.5 inch × 2 inch block. Using the Hole Wizard, add a hole counterbored for a 1/2 inch socket head cap screw with a depth of 2 inches (see **Figure P6.1A**). Use a linear pattern to create a pattern of four evenly spaced holes in the block with a distance of 1.5 inches between hole centers. The resulting block is shown in **Figure P6.1B**, and a section view is shown in **Figure P6.1C**.

FIGURE P6.1A

FIGURE P6.1B

FIGURE P6.1C

P6.2   Using the parts created in Chapter 6, create a working assembly model of a single hinge (**Figure P6.2**).

FIGURE P6.2

P6.3 Create solid models of a 5-ft. long, 1.5-in. diameter pole (**Figure P6.3A**), and a 4-in. diameter sphere with a 1.5-in. diameter hole in it (**Figure P6.3B** and **Figure P6.3C**). Using these models and the flange model created in Chapter 1 (**Figure P6.3D**), create an assembly model of a flagpole (**Figure P6.3E**).

FIGURE P6.3A

FIGURE P6.3B

FIGURE P6.3C

FIGURE P6.3D

FIGURE P6.3E

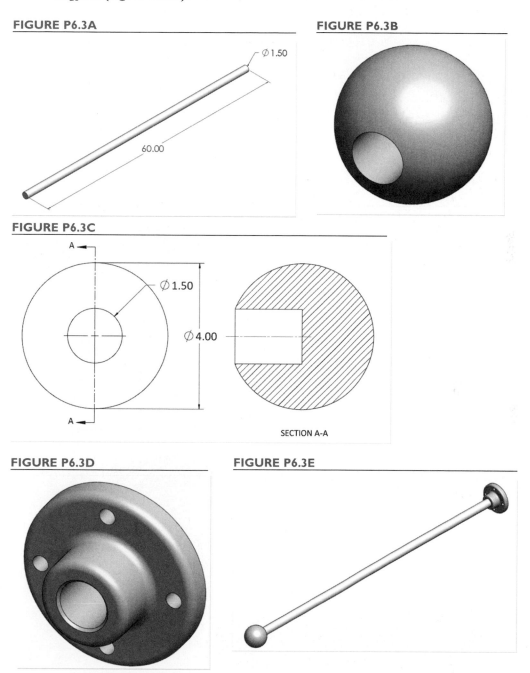

P6.4 Create solid models of the shaft segment shown in **Figure P6.4A** and **Figure P6.4B**, and the key shown in **Figure P6.4C**. Create an assembly with these two parts and the pulley model described in Chapter 1. The finished assembly is shown in exploded state in **Figure P6.4D** and in normal state in **Figure P6.4E**. (Note: We will learn how to create an exploded view in Chapter 7.)

FIGURE P6.4A

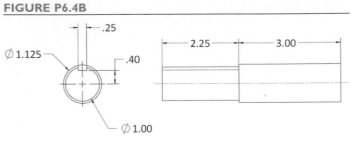

FIGURE P6.4B

FIGURE P6.4C

FIGURE P6.4D

FIGURE P6.4E

P6.5 Create models of the 2 × 4 members shown in **Figure P6.5A**. (Note that 2 × 4s have actual finished dimensions of 1.5 × 3.5 inches.) Create an assembly from these 2 × 4s as shown in **Figure P6.5B**. Consider creating one or more subassemblies to reduce the total number of mates required.

FIGURE P6.5A

FIGURE P6.5B

P6.6 Using simple components (flat boards, 2 × 4s, etc.), design a shelving unit for your room or apartment. Customize your design so that some items that you own (TV, stereo, etc.) will fit on the shelves.

P6.7 The atomic crystal structures of metals are often modeled using spheres to represent atoms. The structure illustrated in **Figure P6.7A** is called a *body-centered cubic* structure. Create a model of this crystal structure as follows:

FIGURE P6.7A

1. Create a 2-inch diameter sphere part model.
2. Open a new assembly.
3. Open a 3-D sketch in the assembly.
4. Add lines to create a box shape, as shown in **Figure P6.7B**. Add relations so that all lines are along the *x*, *y*, and *z* axes. Add an Equal relation to three of the lines so that the sketch defines a cube.
5. Add a centerline connecting two opposite corners of the cube. Add a 4-inch dimension of this line, which will ensure that the atoms on the corners will touch the center atom. Add a point to the midpoint of the centerline, as shown in **Figure P6.7C**.
6. Close the sketch.

FIGURE P6.7B

FIGURE P6.7C

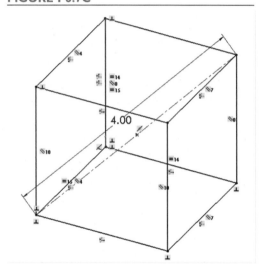

7. Insert nine atoms into the assembly. Turn on the display of the origins.
8. Add mates between the origin of each of the atoms and a corner or the center point of the 3-D sketch, as shown in **Figure P6.7D**.
9. Turn off the display of the origins, and hide the 3-D sketch.

FIGURE P6.7D

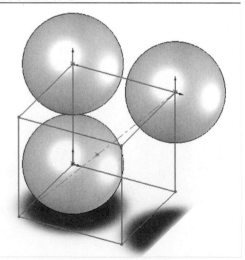

P6.8 The *packing factor* of a crystal structure quantifies how tightly packed the atoms of the structure are, and is defined as the volume of atoms contained within a *unit cell*, or basic building block of the crystal structure, divided by the total volume available within the cell. Find the packing factor of the body-centered cubic structure by following this procedure:

FIGURE P6.8

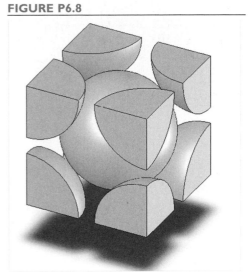

1. Cut away parts of the model created in Problem 6.7 to create a unit cell, as shown in **Figure P6.8**.

2. Use the Mass Properties Tool to find the volume of atoms within the unit cell.

3. Calculate the volume of a solid cube of the same dimensions as the unit cell.

4. Divide the volume of the atoms by the volume of the solid cube.

(Answer: 68%)

P6.9 Repeat Problem 6.7 for the *face-centered cubic* structure in **Figure P6.9A**. The 3-D sketch defining the atom positions is shown in **Figure P6.9B**. Note that a point must be added at the center of each of the six faces of the cube.

FIGURE P6.9A

FIGURE P6.9B

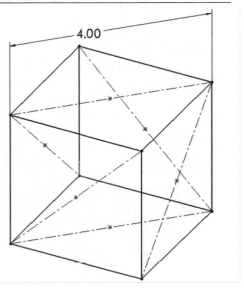

4.00

P6.10 Find the packing factor of the face-centered cubic structure model in P6.9. The unit cell is shown in **Figure P6.10**.

(Answer: 74%)

FIGURE P6.10

P6.11 Create an assembly model of a split hub clamp, as shown in **Figure P6.11A**. To do this, create the part shown in **Figure P6.11B**. Assemble two instances of this part along with two appropriately sized socket-head cap screws to complete the assembly.

FIGURE P6.11A

FIGURE P6.11B

P6.12 Create the assembly model of the cylindrical roller bearing shown in **Figure P6.12A**. To do this, perform the following steps:

FIGURE P6.12A

a. Create the bearing race by revolving the cross-section shown in **Figure P6.12B** (dimensions in inches), and adding .02″ radius fillets to external edges.

b. Create a roller by modeling a 0.4″ long × 0.32″ diameter cylinder, and adding 0.2″ fillets to the edges.

c. Insert the race into a new assembly model and insert a single instance of the roller. Use one coincident and one tangent mate to add the first rolling element, as shown in a Section View in **Figure P6.12C**.

FIGURE P6.12B

```
|------ .700 ------|
 .100
        .100            .160

            R.728
```

FIGURE P6.12C

d. Insert a second instance of the roller. Use one coincident mate and one tangent mate to place the roller in the race and a second tangent mate to place the two rollers in contact (as shown in a Section View in **Figure P6.12D**).

e. Repeat the previous step until twelve rollers have been placed in the model.

FIGURE P6.12D

P6.13 Create a model of an appropriately sized stepped shaft, and assemble the bearing of Problem P6.12 onto the shaft. The assembly should be similar to **Figure P6.13**. There should be no interference in the assembly.

FIGURE P6.13

P6.14 Model the workbench shown in **Figure P6.14**. The bench is constructed from 2 × 4 members (actual dimensions 1.5 by 3.5 inches) and 1-inch-thick plywood for the top and shelf. The top is 72 inches wide by 30 inches deep, and is 34 inches from the floor. The top has a 2-inch overhang on the front and both sides (but not the back). Set the material to Pine for all members, and use the Mass Properties Tool to estimate the weight of the assembled bench.

FIGURE P6.14

CHAPTER 7

Advanced Assembly Operations

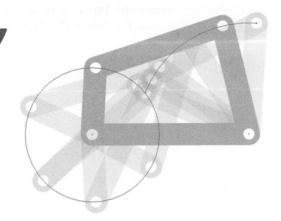

Introduction

In Chapter 6, the basic operations required to create a SolidWorks assembly were introduced. In this chapter, some advanced modeling, visualization, and analysis operations will be developed. The door sub-assembly and part models created in Chapter 6 will be used in the tutorials.

Chapter Objectives

In this chapter, you will:

■ add assembly-level features to an assembly model,

■ create an exploded view of an assembly model,

■ use the Animation Controller to view and save an animation of an exploded view, and

■ use the interference detection and collision detection features to analyze the assembly model.

7.1 Adding Features at the Assembly Level

In Chapter 6, assemblies were created from pre-existing part files. It is sometimes desirable to make modifications to the parts in the assemblies. In this section, the addition of features at the assembly level will be described.

FIGURE 7.1

Open the assembly file Door.SLDASM created in Chapter 6, which is shown in Figure 7.1.

Based on our design intent, we would like to produce a hole in the hatch component corresponding to each of the four countersunk holes in the hinge component. Therefore, we will use the existing holes in the hinge to establish relations that precisely locate our new holes to match the holes in the hinge. The first step will be to create points on the top surface of the hatch where the holes will be added.

Select the top surface of the hatch and choose the **Normal To View**, as shown in Figure 7.2. Select the **Point Tool** from the Sketch group of the CommandManager, as shown in Figure 7.3, and add points at the center of the holes in the hinge, as shown in Figure 7.4. (It may be necessary to hold the cursor momentarily over the perimeter of the hole to "wake up" its center mark.) Close the sketch, as shown in Figure 7.5.

FIGURE 7.2

FIGURE 7.3

FIGURE 7.4

FIGURE 7.5

It will be easier to add the holes in the proper position if the hinge is hidden. Otherwise, the Hole Wizard may not find the proper surface to define the hole.

Right-click on the hinge in the FeatureManager, and select Hide, as shown in Figure 7.6.

The hinge is now hidden from view, as shown in Figure 7.7. The points of the sketch just created should be visible. If they are not, select **View Sketches** from the **Hide/Show Items** menu of the Heads-Up View Toolbar.

FIGURE 7.6

FIGURE 7.7

FIGURE 7.8

Select the Assembly group of the CommandManager. Click on the Assembly Features Tool, which reveals a menu of assembly-level features, as shown in Figure 7.8. Select the Hole Wizard.

Four *pilot holes* will be added to the hatch, matching the positions of the holes in the hinges. (Note: the holes in the hatch are drilled undersized to allow the screws to thread into the hatch.)

In the Hole Specification PropertyManager, with the Type tab selected, set the Hole Type to plain hole, the Size to 7/64 (inches), the End Condition to Blind and the depth to 0.75 inches, as shown in Figure 7.9.

Click on the Positions tab in the PropertyManager, and the Hole Position PropertyManager will prompt you for the position of the holes. Click on each of the points created earlier, as shown in Figure 7.10, and click the check mark twice to create the holes.

Right-click on the hinge in the FeatureManager and select Show.

The holes added match the positions of the holes in the hinge, as shown in Figure 7.11.

Repeat the entire operation on the other hinge.

Right-click on each of the sketches defining the hole positions and select Hide.

Save the changes to the assembly by selecting File: Save.

FIGURE 7.9

FIGURE 7.10

FIGURE 7.11

FIGURE 7.12

7.2 Adding Fasteners to the Assembly

We will now add wood screws to the assembly model. Commercial and educational licenses of SolidWorks have an add-in feature called SmartFasteners that can be used to intelligently insert appropriate mechanical fasteners into an assembly. Some limited-license SolidWorks products do not contain this feature, so we will create a wood screw part model for this exercise.

Open a new part. In the Top Plane, sketch and dimension a 0.190-inch diameter circle, centered at the origin. Extrude the circle upward to a height of 1.00 inches, as shown in Figure 7.12.

DESIGN INTENT | Part-Level and Assembly-Level Features

Since the holes that we are adding in the tutorial affect only the hatch part, we could have added them to the part file. Instead, we have added them at the assembly level, and the holes do not appear in the part model (verify this by opening the hatch file after adding one or more of the holes to the assembly). There are two types of features: *part-level* features, and *assembly-level* features. Where we apply the features should reflect the actual manufacturing and assembly processes. For example, if the holes are predrilled into the hatch before the assembly with the hinges, then the holes should be added to the part model and are considered part-level features. If the holes are added after placing the hinge on the hatch and using the hinge's holes to locate the

drilled holes in the hatch, then these are assembly-level features.

The difference between part-level and assembly-level features is especially important when creating detailed drawings. In this example, the part drawings would contain all information needed to manufacture and/or inspect the hinge and hatch parts. The assembly drawing would show which parts make up the assembly (in a *Bill of Materials*), and would contain only the dimensions necessary to assemble them. In this case, the dimensions to locate the hinges and the hole definitions would be defined on the assembly drawing (we will learn how to make an assembly drawing in Chapter 8).

FIGURE 7.13

Midpoint
Relation

Switch to the Front View. In the Front Plane, sketch and dimension the three lines and vertical centerline shown in Figure 7.13. Select the Revolved Boss/Base Tool from the Features group of the CommandManager, as shown in Figure 7.14. Click the check mark to accept the 360-degree default revolution.

The revolved screw head is shown in **Figure 7.15**.

FIGURE 7.14

FIGURE 7.15

FIGURE 7.16

At the bottom of the screw, sketch and dimension the lines shown in Figure 7.16 in the Front Plane. Add a vertical centerline.

Note that the "cutting tool" profile extends below the bottom of the part. An error is sometimes encountered when the profile comes to a single point on the centerline.

Select the Revolved Cut Tool from the Features group of the CommandManager, as shown in Figure 7.17. Click the check mark to accept the 360-degree default revolution.

The screw body is shown in **Figure 7.18**.

FIGURE 7.17

FIGURE 7.18

Open a sketch on the screw head and sketch a rectangle centered at the origin. The rectangle should extend beyond the edges of the screw by 0.05 inches, and should be 0.055 inches tall, as shown in Figure 7.19. Extrude a cut 0.040 inches deep to create the slot in the screw head, as shown in Figure 7.20. Save this part file as "Screw."

FIGURE 7.19

.055

.050

FIGURE 7.20

Switch back to the door assembly. Expand the definition of the first hinge in the FeatureManager, and click on the first hole created in the hinge, as shown in Figure 7.21. This will highlight the hole, as shown in Figure 7.22.

FIGURE 7.21

```
Door  (Default<Display State-1>)
    Sensors
  ⊞ Annotations
     Front Plane
     Top Plane
     Right Plane
     Origin
  ⊞ (f) Hatch<1>
  ⊟ Hinge<1>
    ⊞ Mates in Door
       Sensors
    ⊞ Annotations
       Material <not specified>
       Front Plane
       Top Plane
       Right Plane
       Origin
    ⊞ Extrude-Thin1
    ⊞ Extrude1
       LPattern1
    ⊞ CSK for #10 Flat Head Machine Screw1
       LPattern2
  ⊞ Hinge<2>
```

FIGURE 7.22

By placing our first screw in the first hole created, we can use the pattern that was previously defined for the holes to place the other screws, rather than placing them one at a time or defining a new pattern.

FIGURE 7.23

Insert a screw into the assembly, placing it approximately in the position shown in Figure 7.23.

Select the Mate Tool. Select the conical surface of the screw head, as shown in Figure 7.24. Select the conical face of the first hole in the hinge, as shown in Figure 7.25.

By default, a concentric mate will be created. However, we can override the default to create a coincident mate. Adding a coincident mate to the two conical surfaces will completely locate the screw in the hole.

FIGURE 7.24

FIGURE 7.25

FIGURE 7.26

In the Mate PropertyManager, change the type of mate to Coincident, as shown in Figure 7.26. Click the check mark to apply the mate, and click the check mark again to close the Mate PropertyManager.

The screw is shown in its final position in Figure 7.27. Rather than add the remaining screws individually, we will create a pattern.

FIGURE 7.27

Since the pattern of the screws will follow that of the holes in the hinges, we will use that pattern to create a *feature-driven* pattern.

From the Assembly group of the CommandManager, click the arrow under the Linear Component Pattern Tool and select Feature Driven Component Pattern, as shown in Figure 7.28. Select the screw as the component to be patterned. For the Driving Feature, click on one of the holes in the hinges, as shown in Figure 7.29.

Click the check mark to complete the pattern, which is shown in Figure 7.30.

FIGURE 7.28

FIGURE 7.29

FIGURE 7.30

The screws in their holes are shown in **Figure 7.31** in wireframe mode. Note that the screws are larger in diameter than the pilot holes in the hatch.

Repeat for the other hinge, as shown in Figure 7.32. Save the changes to the assembly.

FIGURE 7.31

FIGURE 7.32

7.3   Creating an Exploded View

Exploded views are often used to visualize assemblies. In this section, an exploded view of the door assembly will be created.

FIGURE 7.33

Select the Exploded View Tool from the Assembly group of the CommandManager, as shown in Figure 7.33.

Click on each of the screws to select them. If you accidentally select one of the hinges, click on it again to cancel its selection.

These components will appear in the Settings box in the PropertyManager. A manipulator handle will appear in the model window, as shown in Figure 7.34. This allows for "drag and drop" explosion of assembly components.

Click and hold on the manipulator handle that points in the Y direction, and drag the fasteners up to the desired location, as shown in Figure 7.35. Release the mouse button to place the components.

FIGURE 7.34

FIGURE 7.35

FIGURE 7.36

Note that this explosion step, denoted Explode Step1, now appears in the Explode Steps box of the PropertyManager. We can make modifications to this step by using the PropertyManager entries.

Double-click on Explode Step1 in the Explode Steps box of the Property-Manager to select it. Change the distance to 4 inches, as shown in Figure 7.36. Click Apply to change the distance, and click Done to end Explode Step1.

The Explode PropertyManager should be open on the screen. The Settings box should be highlighted, prompting for the selection of components to explode.

Click on both of the hinges to select them.

Click and hold on the manipulator handle that points in the Y direction, and drag the hinges up to the desired location, as shown in Figure 7.37. Release the mouse button to place the components.

Double-click on the entry in the Explode Steps box of the PropertyManager to select it. Change the distance to 2 inches. Click Apply to change the distance, and click Done to end the explode step.

FIGURE 7.37

Click the check mark to complete the exploded view. Save the modified assembly.

Now that the exploded view has been defined, you may toggle between the exploded and collapsed views at any time.

In the FeatureManager, right-click on the name of the door assembly. Select Collapse (Figure 7.38).

The exploded view will be toggled to the collapsed view.

Right-click on the name again, and select Explode, as shown in Figure 7.39.

This will toggle the display back to the exploded state.

The assembly steps can still be modified at any time.

From the top of the FeatureManager, select the ConfigurationManager tab (Figure 7.40).

FIGURE 7.38

FIGURE 7.39

FIGURE 7.40

The ConfigurationManager will appear.

Expand the entries in the ConfigurationManager so that the defined Explode Steps will appear.

Right-click on one of the Explode Step entries, and select Edit Feature, as shown in Figure 7.41.

FIGURE 7.41

DESIGN INTENT | Manufacturing Considerations

The placement of the holes in the hatch and the hinges can be accomplished in one of three ways, and each way simulates a different approach to the manufacturing of the door assembly. First, the holes could be added to the hatch at the part level, just as they are in the hinge part. In this method, at the final assembly step, the hatch and the hinges would be received with the holes in place. The hinges would be aligned with the holes in the hatch, and the fasteners inserted. In this case, the holes would be classified as part-level features. In the second method, neither the hinge nor the hatch would have holes before arriving at the assembly step, where the holes would be classified as assembly-level features.

In the third method, the hatch would be received without holes, and the hinges received with the holes pre-drilled. The holes in the hatch would be drilled to match the holes in the hinges. This is the method modeled in the tutorial. In this case, we have combined a part-level feature (the holes in the hinges) with assembly-level features (the holes in the hatch). Although the final assembly would look the same regardless of which method is chosen, the definition of where the holes are added can be an important consideration in the actual manufacturing process, and the method used to create the solid model should represent the actual manufacturing steps.

The Explode PropertyManager will appear. The entries can be modified, and new steps can be added if desired.

Experiment with changes if you desire. Undo any changes you made, and save the assembly, which will be used in the creation of an assembly drawing in Chapter 7.

FIGURE 7.42

Sketch lines can be added to the exploded view to show how the parts fit together. Although they are not really necessary for a simple assembly such as our door, they can be very helpful in more complex assemblies.

Select the Explode Line Sketch Tool from the Assembly group of the CommandManager, as shown in Figure 7.42.

A 3-D sketch will be opened. By default, the Route Line Tool will be active, as shown in Figure 7.43.

FIGURE 7.43

Click on the point or cylindrical surface of one of the fasteners. Then click on the edge of the corresponding hole in the hatch, as shown in Figure 7.44. Click the right mouse button to create the sketch line.

FIGURE 7.44

Repeat for the other fasteners, as shown in Figure 7.45. Close the sketch to complete the operation.

FIGURE 7.45

The completed explode lines are shown in Figure 7.46. The 3-D sketch containing the explode lines is stored in the ConfigurationManager with the associated exploded configuration, as shown in Figure 7.47.

While this exploded configuration offers a clear picture of how the components are assembled together, we may wish to produce an animation file to demonstrate the explosion step by step. Not only can we easily produce this animation using SolidWorks, we can export the animation as a standard video AVI file that can be viewed by anyone with video playback software.

FIGURE 7.46

FIGURE 7.47

Right-click on the Door Configurations(s) entry in the ConfigurationManager, and select Animate Collapse, as shown in Figure 7.48.

The Animation Controller will initiate (shown in Figure 7.49), and the animation of the collapse/explosion will begin. The Animation Controller contains buttons to control the play, playback mode, and play speed of the on-screen animation.

FIGURE 7.48

FIGURE 7.49

Click the Stop button to end the animation, as shown in Figure 7.50. Set the Playback Mode to Normal, as shown in Figure 7.51.

FIGURE 7.50

FIGURE 7.51

FIGURE 7.52

In order to view this animation using the Animation Controller, access to the assembly model, all part models used in the assembly, and the SolidWorks software package are required. To allow someone without access to the models and software to view this animation, we can save it as a video AVI file.

In the Animation Controller, click the Save Animation button, as shown in Figure 7.52.

FIGURE 7.53

The Save Animation to File dialog box will appear, as shown in **Figure 7.53**. This allows us to select the appropriate directory and filename for a video file. It also allows us to set some parameters that control file size and image quality, such as the number of frames per second to be saved when we create our video file. For this application, we will accept the default values.

Browse to the appropriate file location, and click Save.

The Video Compression dialog box will appear, as shown in **Figure 7.54**. This dialog box allows us to select the video resolution of the video file. Higher-resolution images will result in larger file sizes. For this application, we will accept the default values for Compressor type and quality.

Click OK to close the Video Compression dialog box.

FIGURE 7.54

The animation will be replayed in the graphics window. While this is occurring, the video file will be written at the frames per second and compression level we specified.

The video file will be stored using the name and location we specified. In this case, the file will be name Door.avi. This file is a stand-alone video file; it is not linked in any way to your SolidWorks part or assembly files. As such, it can be viewed by anyone with standard video playback software, such as Windows Media Player or RealPlayer. Do note, however, that all associativity with the model is lost; subsequent changes to the assembly in SolidWorks will not affect this video file. If changes are desired, a new video file must be created.

Close the assembly window, without saving the last changes made.

7.4 Detecting Interferences and Collisions

In the assembly mode, the SolidWorks program has the ability to check for interferences between components. It does this by determining locations where solid volumes overlap. This tutorial will demonstrate these capabilities.

Open the file "hatch assembly.SLDASM" created in Chapter 6. Click and drag on the hatch to rotate it into the approximate position shown in Figure 7.55. Press the Esc key to deselect the hatch.

In doing this, we have intentionally introduced an interference between the door and the frame.

From the main menu, select Tools: Interference Detection.

The PropertyManager shows the parts for which interference is to be analyzed. By default, the entire assembly is selected.

Click the Calculate button to commence interference detection (Figure 7.56).

In the PropertyManager, the interference between the frame and hatch is identified, as shown in Figure 7.57. Note that there are nine regions of interference detected. In addition to the overlap of the door and the frame, the eight screws in their undersized holes are detected.

FIGURE 7.55

FIGURE 7.56

FIGURE 7.57

Click the X in the PropertyManager to end interference detection.

Although we have seen that components in an assembly can be moved by simply clicking and dragging them, the Move Component Tool allows collision detection to be incorporated into part movements.

Click and drag the hatch so that it no longer overlaps the frame.

Select the Move Component Tool from the Assembly group of the CommandManager, as shown in Figure 7.58. In the

FIGURE 7.58

FIGURE 7.59

PropertyManager, check "Collision Detection" and "Stop at collision." Under Advanced Options, select "Highlight faces" and "Sound," as shown in Figure 7.59.

Move the hatch toward the frame until it stops, as shown in Figure 7.60.

Note that when the hatch and frame faces touch, the faces are highlighted and a tone signifies a collision.

Now drag the hatch upward as far as possible, as shown in Figure 7.61.

The movement stops when the hinges collide, as shown in the Right View of Figure 7.62.

Click the check mark or hit the Esc key to deselect the Move Component Tool, and close the file.

FIGURE 7.60   FIGURE 7.61

FIGURE 7.62

PROBLEMS

P7.1 Import the part model created in Problem P6.1 into a new assembly. Import a socket-head cap screw from Chapter 4 into the assembly. Right-click on the cap screw and select properties. Select the 101 configuration. Add mates to locate the cap screw, and define a linear pattern to place cap screws in the other holes.

FIGURE P7.1

P7.2 Create an exploded view of the assembly created in **Figure P7.1**.

FIGURE P7.2

P7.3 Create an assembly from the frame part and two hinges, as shown in **Figure P7.3A**. The hinges should be located as shown in **Figure P7.3B**. Add 7/64 holes (0.75 inch deep) to the frame, aligned with the holes in the hinges, and insert screws to attach the hinges to the frame.

FIGURE P7.3A   **FIGURE P7.3B**

P7.4   Create a new hatch assembly, using the door assembly from Chapter 7, the frame/hinge assembly from P7.3, and two pins.

FIGURE P7.4

P7.5   Create an exploded view of the flagpole assembly model created in Problem P6.3.

FIGURE P7.5

P7.6   Create an exploded view of the hinge model created in Problem P6.2.

FIGURE P7.6

P7.7   Add an angular dimension between the faces of the hinges in the hinge assembly model, as shown in **Figure P7.7**. Use the Move Component Tool with collision detection to move the hinge into its limiting positions. What is the total angle through which the hinge can be rotated?

(Answer: 284 degrees)

FIGURE P7.7

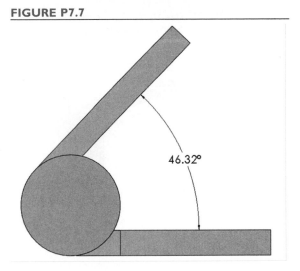

46.32°

P7.8   Create an exploded view of the pulley assembly model created in Problem P6.4.

FIGURE P7.8

P7.9   Create and export an animation (*.avi file) of the explosion from Problem P7.8.

P7.10 Open the flagpole assembly created in Problem P6.3. Modify the diameter of the pole component to 2.00 inches, and rebuild the assembly. Use the Interference Detection Tool to locate the interferences between the components in the rebuilt assembly.

P7.11 Create an exploded view of the split hub clamp assembly model created in Problem P6.11.

FIGURE P7.11

FIGURE P7.11

P7.12 Create and export an animation (*.avi file) of the explosion from Problem P7.11.

P7.13 Create an exploded view of the stepped shaft assembly model created in Problem P6.13.

FIGURE P7.13

CHAPTER 8

Creating Assembly Drawings

Introduction

In Chapters 6 and 7, the development of solid models of complex assemblies from SolidWorks part files was described. In this chapter, the documentation of these assemblies through the use of 2-D assembly drawings will be introduced.

8.1 Creating an Assembly Drawing

In this section, an assembly drawing of the door assembly created in Chapter 6 and modified in Chapter 7 will be produced.

Open the assembly file "Door.SLDASM" (modified in Chapter 7). Make sure that the assembly is in the "collapsed" configuration. Select File: New from the main menu, and choose Drawing from the template choices.

Select the sheet format that you created in Chapter 2, as shown in Figure 8.1, if desired. If you prefer a blank drawing sheet, select A-Landscape and clear the "Display sheet format" box.

FIGURE 8.1

Chapter Objectives

In this chapter, you will:

- create a 2-D assembly drawing,

- incorporate an exploded view into an assembly drawing,

- generate a Bill of Materials, and

- explore the associativity between part models, assembly models, and assembly drawings.

Select the Standard 3 View Tool from the View Layout group of the CommandManager, as shown in Figure 8.2. In the PropertyManager click on the assembly name ("Door") to select it. Click the check mark.

Three standard drawing views will be displayed, as shown in **Figure 8.3**.

FIGURE 8.2

FIGURE 8.3

Introduction to Solid Modeling
Using SolidWorks

If your drawing views display hidden lines, click on the Front View and select the Hidden Lines Removed Style from the Display Style menu of the Heads-Up View toolbar, as shown in Figure 8.4. Also, click on each centerline, as shown in Figure 8.5, and delete them.

Since an assembly drawing's purpose is to show how components are joined, hidden edges and centerlines are not usually displayed.

FIGURE 8.4

Hidden Lines Removed
Displays only those model edges that can be seen from the current view orientation.

FIGURE 8.5

Right-click in the blank area of the drawing and select Properties. Change the scale to 1:5, as shown in Figure 8.6.

FIGURE 8.6

Sheet Properties

Name: Sheet1   Type of projection
                      ○ First angle
Scale: 1 : 5   ● Third angle

Sheet Format/Size
● Standard sheet size
A1 - Landscape
A2 - Landscape   Reload

FIGURE 8.7

Model Items
Imports dimensions, annotations, and reference geometry from the referenced model into the selected view.

We will now add dimensions to the drawing. For an assembly drawing, we want to show only the dimensions associated with assembly-level features and operations.

Select the Top View. From the Annotation group of the CommandManager, select Model Items, as shown in Figure 8.7.

In the Model Items PropertyManager, select "Only assembly" from the Source menu, as shown in Figure 8.8. Click the check mark to complete the operation.

The dimensions related to the placement of the hinges are imported, as shown in **Figure 8.9**. You may need to drag them into the positions shown.

FIGURE 8.8

FIGURE 8.9

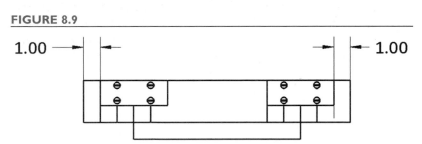

Although it is not necessary to show both dimensions, we will see later that we can adjust the locations of the hinges by changing these dimensions on the drawing. Since the two hinges were placed in the assembly separately, the two 1-inch dimensions are not related. Note that if we wanted to add an equation to the assembly model, then only one of the dimensions would be needed.

8.2 Adding an Exploded View

In Chapter 7, an exploded view of the door assembly was created. This exploded view can be easily added to the drawing.

Select the Model View Tool from the View Layout group of the CommandManager, as shown in Figure 8.10.

The door assembly should be selected by default in the PropertyManager. If it is not, select it. Click the Next arrow, as shown in Figure 8.11.

FIGURE 8.10

FIGURE 8.11

DESIGN INTENT | Assembly-Level Dimensions

When importing dimensions into our assembly drawing, we selected the option labeled "Only assembly," as opposed to "Entire model." As a result, we imported dimensions associated only with assembly-level features. If the other option had been chosen, then the dimensions defining the components would have been imported as well. These dimensions could be edited, resulting in changes to the part files.

Including component dimensions in an assembly drawing is not usually recommended. One reason is that the components are defined in separate drawings, so adding the dimensions to the assembly drawings is redundant. Another reason is that a component may be used in multiple assemblies, so editing a component at the assembly level may produce unexpected changes to other assemblies.

FIGURE 8.12

The PropertyManager will now contain a list of available orientations associated with the door assembly.

Select Single View, and check the Trimetric View, as shown in Figure 8.12. **Scroll down in the PropertyManager, and select Wireframe (without hidden lines) as the display type, and check the "Use sheet scale" option, as shown in** Figure 8.13.

Click on the location in the drawing window where the Trimetric View will appear, as shown in Figure 8.14.

With the new view selected, select More Properties from the PropertyManager.

FIGURE 8.13

FIGURE 8.14

Introduction to Solid Modeling
Using SolidWorks

The Drawing View Properties dialog box will appear.

Check the "Show in exploded state" box to change the currently selected view to an exploded view (Figure 8.15), and click OK.

The exploded view will now appear in the drawing window, as shown in **Figure 8.16**.

FIGURE 8.15

FIGURE 8.16

We will now add a note regarding the placement of the holes in the hatch.

Select the Note Tool from the Annotation group of the Command-Manager, as shown in Figure 8.17.

Click on the edge of one of the holes in the exploded view. This will create a leader for the note, as shown in Figure 8.18.

Click on the approximate location of the note, as shown in Figure 8.19.

FIGURE 8.17

FIGURE 8.18

FIGURE 8.19

FIGURE 8.20

If desired, change the font type and size. Enter the text shown in Figure 8.20.

Click in the drawing window to place the note, and then press the Esc key to end the Note command. Click and drag the note to its final position.

The drawing is shown in Figure 8.21.

FIGURE 8.21

8.3   Creating a Bill of Materials

It is often desirable to generate a parts list, or Bill of Materials (BOM), associated with an assembly. The SolidWorks program can automatically create this list from an assembly file.

Select any of the drawing views. From the Annotation group of the CommandManager, select Tables: Bill of Materials, as shown in Figure 8.22.

FIGURE 8.22

Accept the default selections listed in the PropertyManager, and click the check mark (Figure 8.23).

Click anywhere in the drawing to place the Bill of Materials, as shown in Figure 8.24.

FIGURE 8.23

FIGURE 8.24

The table can be edited much like an Excel spreadsheet. The formatting of your Bill of Materials may differ slightly, but the following steps in general can be used to control appearance.

Click anywhere in the table to open it for editing.

We will now make changes to the appearance and formatting of the table.

Click on Column Heading C to select the Description column, as shown in Figure 8.25. Click the Hide/Show button to hide the column, as shown in Figure 8.26. Press the Esc key to turn off the tool.

FIGURE 8.25

Click on the arrows in the upper left corner of the table. This causes the entire table to be selected. Click the "Use document font" icon, as shown in Figure 8.27, to override the default font. Select a new font type, if desired, and a larger size, as shown in Figure 8.28.

FIGURE 8.26

Change the width of any column by clicking and dragging on the right boundary of the column, as shown in Figure 8.29.

FIGURE 8.27

FIGURE 8.28

FIGURE 8.29

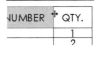

Individual row heights can be changed in a similar manner; however, we usually want the rows to be the same height. Therefore, we will specify the height for all rows.

FIGURE 8.30

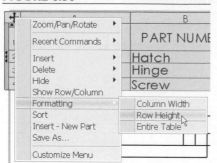

Right-click the arrows in the upper left corner of the table. Select Formatting: Row Height, as shown in Figure 8.30. Set the row height to 0.30 inches, as shown in Figure 8.31.

Click on the arrows in the upper left corner to select the entire table. Click the Center Align tool, as shown in Figure 8.32.

FIGURE 8.31

FIGURE 8.32

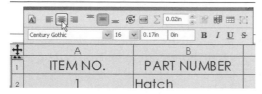

By default, one of the columns is labeled "PART NUMBER." This column contains the file name for each component, as most companies store part files by number rather than the common names that we use in this book. We can easily change the column heading to reflect our method of naming parts.

FIGURE 8.33

Double-click the cell containing the label "PART NUMBER." Change the text to "PART NAME," as shown in Figure 8.33. Click outside the table to accept the change.

Click and drag the move icon in the upper left corner of the Bill of Materials to the position desired.

Our drawing is almost complete, but the item numbers in the Bill of Materials are not linked with the components in the drawing. We will add "balloons" with part numbers to the drawing.

FIGURE 8.34

Select the exploded Trimetric View. Select the AutoBalloon Tool from the Annotation group of the Command-Manager, as shown in Figure 8.34.

Balloons will be added to the view, as shown in **Figure 8.35**. The appearance of the balloons can be changed from the PropertyManager.

FIGURE 8.35

Drill 7/64 holes .75 deep to match holes in hinges

FIGURE 8.36

In the PropertyManager, select I Character from the Size pull-down menu, as shown in **Figure 8.36**. Click the check mark to close the PropertyManager.

Click and drag the part numbers to relocate the balloons as desired.

You can also move the arrow end of the leader. Note that if the leader is attached to an edge or a point, an arrow appears. If the leader is attached to a face, a dot appears at the end of the leader.

Move the drawing views to the desired locations, and add the title with the Note Tool from the Annotation group of the CommandManager. The completed drawing is shown in **Figure 8.37**.

FIGURE 8.37

ITEM NO.	PART NAME	QTY.
1	Hatch	1
2	Hinge	2
3	Screw	8

Drill 7/64 holes .75 deep to match holes in hinges

Introduction to Solid Modeling Using SolidWorks

DOOR ASSEMBLY

Save the drawing with the file name "Door," and close it.

Note that the drawing file has the extension .SLDDRW.

8.4   Investigating Associativity

As with part drawings, assembly drawings maintain *associativity* with the assembly models used to create them; the drawing dimensions are linked to the model dimensions. Changes to assembly drawings will propagate to the assembly models; however, since only assembly-level dimensions are shown on assembly drawings, the part files will be unchanged. On the other hand, since assembly models are composed of part models, changes to part files will propagate to the assembly level and thus to the assembly drawing. In this section, this associativity will be demonstrated.

Open the door assembly file (Door.SLDASM), the hatch part file (Hatch.SLDPRT), and the door drawing file (Door.SLDDRW). From the main menu, select Windows: Tile Vertically to display all three windows at once, as shown in Figure 8.38.

In each window, right-click in the graphics area and select Zoom to Fit to display the entire model. You can also click and drag the FeatureManager boundary in each window to maximize the size of the graphics area.

FIGURE 8.38

In the drawing window, zoom in on the Top View. Double-click on each of the 1-inch dimensions locating the hinges, and change the values to 3 inches. Click the Rebuild Tool, and the hinges will be displayed in the positions shown in Figure 8.39.

FIGURE 8.39

Click in the assembly window, and the assembly will be updated, as shown in Figure 8.40.

If you click in the part window, the hatch part will be unchanged.

We will change the hinge location dimensions back to 1 inch from the assembly window.

In the assembly window, right-click on Annotations in the FeatureManager, as shown in Figure 8.41. Click on Show Feature Dimensions, and the dimensions will be displayed.

Double-click on each of the 3-inch dimensions and change them to 1 inch each, as shown in Figure 8.42. Rebuild the assembly.

If you click in the drawing window, you will see the drawing update to the new dimensions.

FIGURE 8.40

FIGURE 8.41

FIGURE 8.42

We will now see how a change to a part propagates to the assembly and the assembly drawing.

FIGURE 8.43

In the part window, double-click Extrude2 in the Feature-Manager, as shown in Figure 8.43.

The dimensions associated with this feature (the raised portion of the hatch) will be displayed.

Double-click on the 1-inch dimension, and change it to 3 inches. Rebuild the model.

Click in the assembly window. The message shown in Figure 8.44 will appear, letting you know that a component of the assembly has been changed. Click Yes to rebuild the assembly, or wait for the timer to expire.

FIGURE 8.44

The result is shown in **Figure 8.45**.

FIGURE 8.45

Click in the drawing window.

The assembly drawing will show the modified component, as shown in **Figure 8.46** (shown full screen here).

Change the dimension back to I inch in the part window.

We will now change the name of the hatch component. Suppose that there are several other hatch sizes available, and so we want to give the 16-inch hatch a distinctive name.

FIGURE 8.46

ITEM NO.	PART NAME	QTY.
1	Hatch	1
2	Hinge	2
3	Screw	8

Drill 7/64 holes .75 deep to match holes in hinges

Introduction to Solid Modeling Using SolidWorks

DOOR ASSEMBLY

In the part window, select File: Save As from the main menu. The message displayed in Figure 8.47 will be displayed.

FIGURE 8.47

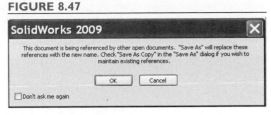

This message is warning you that the part you are about to rename is used in assemblies. If you are creating a copy to use elsewhere, then you will want to use the "Save as copy" option when saving the file. This will maintain the old file as the one that is used in the existing assemblies. If the "Save as copy" box is left unchecked, then the newly created file (with the new file name) will replace the old file in the assemblies. We will use the latter option, since we want the new file name to be used in our assembly.

Click OK to close the message box. In the Save As box, save the file with the file name Hatch16. Leave the "Save as copy" box unchecked. Click Save.

In the assembly window, notice that the new file name is referenced in the FeatureManager (Figure 8.48).

Switch to the drawing window, and rebuild the drawing.

Note that the part name of the hatch has changed in the Bill of Materials (Figure 8.49).

Close all of the windows. It is not necessary to save the changes.

FIGURE 8.48

FIGURE 8.49

ITEM NO.	PART NAME	QTY.
1	Hatch16	1
2	Hinge	2
3	Screw	8

P8.1   Create an assembly drawing, complete with exploded view and bill of materials, for the hinge assembly created in Problem P6.2.

FIGURE P8.1

P8.2   Create an assembly drawing, complete with exploded view and bill of materials, for the flagpole assembly created in Problem P6.3.

FIGURE P8.2

P8.3   Create an assembly drawing, complete with exploded view and bill of
materials, for the shaft assembly created in Problem P6.4.

FIGURE P8.3

P8.4   Create an assembly drawing, complete with bill of materials, for the
frame assembly created in Problem P6.5.

FIGURE P8.4

P8.5   Create an assembly drawing, complete with exploded view and bill of materials, for the frame-and-hinge assembly created in Problem P7.3.

FIGURE P8.5

P8.6   Create an assembly drawing, complete with exploded view and bill of materials, for the split hub clamp model created in Problem P6.11.

FIGURE P8.6

P8.7   Create an assembly drawing, complete with exploded view and bill of materials, for the workbench model created in Problem P6.14.

FIGURE P8.7

PART TWO

Applications of SolidWorks®

CHAPTER 9

Using SolidWorks for the Generation of 2-D Layouts

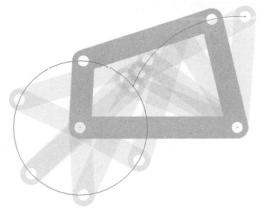

Introduction

Earlier we learned to create 2-D drawings from solid parts and assemblies. For some tasks, however, working directly in the 2-D environment is preferred. For example, floor plans and site drawings, plant equipment layouts, and electrical schematic drawings are usually created in 2-D. The SolidWorks program can be used for these applications, and the ease of changing dimensions allows multiple configurations to be quickly evaluated.

9.1 A Simple Floor Plan Layout

In this exercise, we will prepare a layout drawing of a simple quality assurance lab for a manufacturing shop. The following items need to be placed in the lab:

- A tensile test machine, with a rectangular "footprint" of 4 feet wide by 3 feet deep
- An inspection bench, 8 feet by 3 feet
- A desk, 5 feet by 30 inches
- Three cabinets for storing measuring tools and fixtures, each 4 feet wide by 2 feet deep, with doors that swing outward

Chapter Objectives

In this chapter, you will:

- choose and set the scale for a 2-D layout,

- use a unit system of feet and inches,

- create a floor plan and learn how to modify the layout to evaluate alternate configurations,

- expand your knowledge of using relations with 2-D entities, and

- calculate the areas, centroid locations, and moments of inertias of 2-D areas.

FIGURE 9.1

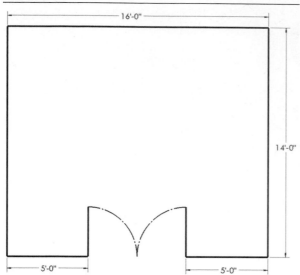

Parts will be brought in bins that are 4 feet square, so there will need to be room for one parts bin. The room that has been proposed for the lab is shown in **Figure 9.1**. We would like to prepare a layout drawing to show how the equipment should be placed in the room.

Begin by opening a new drawing in SolidWorks. When prompted to select a sheet format, pick A-Landscape as the paper size, and make sure the "Display sheet format" box is unchecked, as shown in Figure 9.2. Click OK. If the Model View dialog appears in the PropertyManager, click the x to end the Model View command.

When we import parts into drawings, an appropriate scale is automatically set. When preparing a 2-D layout, we must specify the scale. To make the room fill most of the paper, we can set the scale at 1 inch equals 2 feet. Thus, the 14-foot dimension will appear as 7 inches on the drawing.

FIGURE 9.2

Right-click anywhere in the drawing area, and select Properties. Set the scale to 1:24, as shown in Figure 9.3. Click OK.

FIGURE 9.3

We can use units of feet, inches, or mixed feet and inches for this layout. We will use mixed feet and inches.

Select Tools: Options from the menu. Under the Document Properties tab, click on Units. Select Custom as the unit system, and feet & inches from the pull-down menu of length units. Set the decimal places to None, as shown in Figure 9.4. Click OK.

Select the Corner Rectangle Tool from the Sketch group of the CommandManager, and drag out a rectangle. Dimension the sides of the rectangle, as shown in Figure 9.5. When entering the dimensions as feet, include the foot symbol (') after the number. Otherwise, the dimension units default to inches.

If sketch relation icons do not appear on the drawing, select View: Sketch Relations from the main menu.

FIGURE 9.4

FIGURE 9.5

FIGURE 9.6

With the Dimension Tool turned off, click and drag one corner of the rectangle until the shape is approximately centered on the sheet. Then select Fix in the PropertyManager to fix the point on the sheet, as shown in Figure 9.6.

Delete the bottom line and replace it with the two lines shown in Figure 9.7. Dimension each of the new lines, and add a collinear relation to the lines.

FIGURE 9.7

The drawing should be fully defined.

There are several ways that the doors and the arcs that represent their swing paths can be drawn and dimensioned. The method here uses relations to define the geometry.

FIGURE 9.8

Draw the vertical lines as shown in Figure 9.8.

Select the Centerpoint Arc Tool from the Sketch group of the CommandManager, as shown in Figure 9.9.

This tool allows you to construct an arc by picking the center of the arc and then the two endpoints.

FIGURE 9.9

FIGURE 9.10

5'-0"

Move the cursor directly over the intersection of one of the vertical lines and the adjacent horizontal line, as shown in Figure 9.10. Click once on the intersection to set the center point of the arc. Click again on the other end of the vertical line, as shown in Figure 9.11, to set the starting point of the arc. Now drag an arc approximately 90 degrees, as shown in Figure 9.12, and click to finish the construction of the arc.

Check the "For construction" box in the PropertyManager.

FIGURE 9.11

FIGURE 9.12

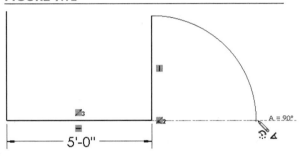

Repeat for the other door, as shown in Figure 9.13.

Select the endpoints of the arcs, as shown in Figure 9.14.

FIGURE 9.13

FIGURE 9.14

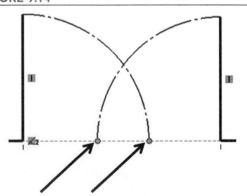

In the **PropertyManager, click Merge, which joins the two points, as shown in** Figure 9.15. **Select both arcs, and add Equal and Tangent relations. The arcs will now appear as shown in** Figure 9.16.

FIGURE 9.15

FIGURE 9.16

The drawing should once again be fully defined.

In order to make the outer walls and doors stand out, we will make those lines thicker. For this, we will use the Line Format toolbar, as shown in Figure 9.17.

FIGURE 9.17

If the Line Format toolbar is not shown, right-click in the Menu Bar or CommandManager. Click on Line Format to add the toolbar.

Select the lines representing the walls and doors. In the Line Format toolbar, select the Line Thickness Tool, as shown in Figure 9.17, **and pick a thicker line than the default. Click the check mark in the PropertyManager to apply the selected line thickness.**

FIGURE 9.18

The walls and doors now stand out from the dimension and construction lines, as shown in Figure 9.18.

We will now place the tensile test machine in the room.

Draw a rectangle somewhere in the room, and dimension it as 4 feet by 3 feet, as shown in Figure 9.19.

FIGURE 9.19

The dimensions that we added are not necessary to show in the drawing, but are required to set the size of the machine's footprint. We can hide the dimensions so they do not show, but still control the size of the rectangle.

Turn off the Smart Dimension Tool. Right-click on one of the dimensions and select Hide, as shown in Figure 9.20. Repeat for the other dimension.

FIGURE 9.20

The rectangle is fully defined except for its position on the drawing. By clicking and dragging on one of the corners, you can move it.

Click and drag a corner of the rectangle until it is placed near the upper-right corner of the room, as shown in Figure 9.21.

Note: if you drag a corner until it contacts another point or a line, as shown in **Figure 9.22**, then a relation is automatically created with that entity, and you will be unable to move the rectangle away from the entity. If you desire to do so, then click on the coincident icon and delete it.

FIGURE 9.21   **FIGURE 9.22**

While it is not necessary to fix the rectangle in place, doing so is a good idea in that it will make the drawing fully defined if all other required dimensions and relations are present. If you fix the position of the rectangle and the status bar reports that the drawing is still underdefined, then you will probably want to check existing dimensions and relations to see why.

Click on a corner of the rectangle to select that point and display its properties in the PropertyManager. Click Fix to fix that point, and close the PropertyManager. Press Esc to deselect the point.

FIGURE 9.23

The drawing should now be fully defined. We will now label the rectangle as the location of the tensile test machine.

Click on the Annotation tab of the CommandManager. Select the Note Tool, as shown in Figure 9.23. Click at the approximate location of the note, near the new rectangle.

Set the desired font, size, and alignment (centered), as shown in Figure 9.24. Type "Tensile Test Machine" in the text box.

FIGURE 9.24

Click anywhere outside of the note box, and then Esc to end the note creation.

(If you want to place the same note in another location, then you can click the note down in multiple locations before using the Esc key to end the process.)

Click and drag the note to its final position, as shown in Figure 9.25.

If you want the note font changed for the entire drawing, do so from Tools: Options: Document Properties: Annotations from the main menu, as shown in **Figure 9.26**.

FIGURE 9.25

FIGURE 9.26

FIGURE 9.27

FIGURE 9.28

Add the first of the cabinets by drawing and dimensioning the rectangle as shown in Figure 9.27. Hide the dimensions.

After drawing the cabinet and moving its location around, you may decide that turning the cabinet 90 degrees will allow it to fit into the space better. The easiest way to do this is simply to switch the values of the dimensions. This will require showing and editing the currently hidden dimensions.

Select View: Hide/Show Annotations from the main menu, as shown in Figure 9.28. The hidden dimensions will be shown in gray. Click on each dimension that you want to show, as shown in Figure 9.29. Select Hide/Show Annotations again or Esc to return to the editing mode. Change the cabinet dimensions as shown in Figure 9.30. Hide the dimensions again.

FIGURE 9.29

FIGURE 9.30

FIGURE 9.31

Add a note identifying the cabinet. If desired, rotate the text 90 degrees by clearing the "Use document font" box and changing the angle in the PropertyManager, as shown in Figure 9.31. Click the check mark.

Add the other two cabinets. Dimension, locate, and label them, as shown in Figure **9.32.**

FIGURE 9.32

Note that if you want to use Copy and Paste to add the cabinets, then the dimensions defining their sizes must also be copied, or else the cabinets will change size when dragged into a new position. Hidden dimensions cannot be copied, so you must first unhide the dimensions before using Copy and Paste commands.

Add and dimension rectangles representing the inspection bench and desk, as shown in Figure **9.33.**

FIGURE 9.33

Label the inspection bench and desk. Add and label the 4-foot-square parts bin, with the lines representing it shown as construction lines. Fix all of the entities in position so that the drawing is fully defined. Hide the dimensions defining the bench, desk, and parts bin. The finished drawing is shown in Figure 9.34. Note that the sketch relation icons are not shown when the drawing is printed.

Suppose that we wish to rotate the tensile test machine 45 degrees to give the operator better access to the machine. To do so, it will be necessary to delete some of the relations created automatically when the rectangle was drawn, and to add some new relations to maintain the rectangular shape and orient the rectangle properly.

Click on the Fixed icon, as shown in Figure 9.35. Delete it. Delete the horizontal and vertical relations from the four lines defining the machine position. Add parallel relations between each of the pairs of opposite sides, as shown in Figure 9.36. Select two adjacent sides, and add a perpendicular relation, as shown in Figure 9.37.

The addition of these relations defines the shape as a rectangle, but the removal of the horizontal and vertical relations allows it to be rotated. The hidden dimensions still apply to the rectangle, and fix its size.

FIGURE 9.34

FIGURE 9.35

FIGURE 9.36

FIGURE 9.37

FIGURE 9.38

FIGURE 9.39

Select the Smart Dimension Tool. Select one of the vertical lines of the rectangle and then a horizontal wall, as shown in Figure 9.38.

This will create an angular dimension between the lines, as shown in Figure 9.39. Change the dimensions to 135 degrees, reflecting a rotation of 45 degrees, as shown in Figure 9.40.

Click to select the note. Rotate the text labeling the machine −45 degrees, as shown in Figure 9.41.

Hide the angular dimension, drag it into position, and fix one of the corner points.

The drawing is now complete, as shown in Figure 9.42. Dimensions can be added to precisely place the objects in the room, if required. (It will be necessary to delete the fixed relations of the points if these dimensions are to be added.)

FIGURE 9.40

FIGURE 9.41

FIGURE 9.42

If you would like to establish such dimensions, or further optimize the layout by moving the fixtures and furniture, you can create a block entity that includes both the object shape and the text.

Delete the Fixed relation setting the location of the Inspection Bench, so that both it and its label are free to move. With the Ctrl key depressed, select the four lines and the text that comprise the Inspection Bench. From the main menu, select

FUTURE STUDY

Industrial Engineering

In this example, we created a floor plan to see if the proposed lab space would accommodate the equipment required for the lab. Consider the challenge of planning the layout of a manufacturing plant with hundreds of thousands of square feet of floor space and hundreds of machines. If not well planned, the operations of the plant will be crippled by inefficiencies in the ways that raw materials and parts move through the plant. Efficient plant layout is one of the functions of industrial engineers.

Industrial engineers perform many other functions toward the goal of improving operations. These functions might include monitoring and improving the quality of finished products, streamlining material handling and product flow, or redesigning work cells for better efficiency.

While the word "industrial" refers to the manufacturing environment where industrial engineers have traditionally worked, the skills of industrial engineers are now being widely used in service sector businesses as well. For example, many hospitals use industrial engineers to help improve quality and efficiency. Package-delivery companies, facing monumental logistics challenges associated with delivering packages worldwide under extreme time pressure, also use the services of industrial engineers.

Industrial engineers also are involved with product design, usually from the standpoint of *ergonomics*, the consideration of human characteristics and limitations in the design of products. In this area, there is overlap with the functions of the industrial designer.

Most engineering students in other disciplines take some coursework in industrial engineering topics. These topics could include engineering economy, quality control, project management, and ergonomic design.

Tools:Block:Make, as shown in Figure 9.43. Click the check mark to create the block and close the Make Block PropertyManager.

FIGURE 9.43

A block is a grouping of drawing entities that can now be treated as a single entity. We can now move or rotate the group of entities together.

FIGURE 9.44

Inspection Bench

Click and hold a corner of the Inspection Bench, and drag it to a new location, as shown in Figure 9.44. Note that the entire block (lines and text) move together. If you later decide to separate the entities, the block must be exploded. This can be done by clicking on any part of the block to select it and selecting Tools:Block:Explode from the main menu.

Locate the Inspection Bench as desired, and reapply a fixed relation to locate the block. Save the drawing if desired, and close it.

9.2   Finding the Properties of 2-D Shapes

The SolidWorks program can be used to determine the properties of areas. For example, the areas of complex shapes can be determined, and the locations of *centroids* and the values of *moments of inertia* of areas can be computed.

9.2.1 Calculating the Area of a Shape

FIGURE 9.45

Consider the building lot shown in **Figure 9.45**. Suppose that we want to find the acreage of this lot. The first step is to draw the lot. In order to fit the drawing on an 8-1/2 × 11 inch sheet, the minimum scale factor possible will be 1320 inches (110 feet) divided by 8.5 inches = 155. We will use a scale factor of 1:200.

Open a new drawing. Choose an A-Landscape paper size without a sheet format displayed. Set the drawing scale to 1:200, the units to feet & inches, and draw and dimension the lot shown in Figure 9.45. Fix one corner of the lot, and the drawing should be fully defined.

FIGURE 9.46

FIGURE 9.47

Press the Esc key to deselect any points or lines. Select Tools: Section Properties from the main menu, as shown in Figure 9.46. Click the Options button, and set the units to feet and the number of decimal places to 6, as shown in Figure 9.47. Click OK.

The area will be displayed in the units selected, in this case square feet. However, the area displayed will be that of the *drawing* area, not the actual area. Since the area of the drawing in square feet will be small, a large number of decimal places will be needed for accuracy.

The area displayed is 0.318910 square feet, as shown in **Figure 9.48**. To determine the area of the lot, the calculated area must be adjusted to account for the scale. Since area has the units of length squared, the calculated area must be multiplied by the scale factor squared:

Area = (0.318910 ft^2)(200^2) = 12,756 ft^2

To convert to acres, the conversion factor of 43,560 ft^2/acre must be applied:

Area = (12,756 ft^2)/(43,560 ft^2/acre) = 0.29 acres

FIGURE 9.48

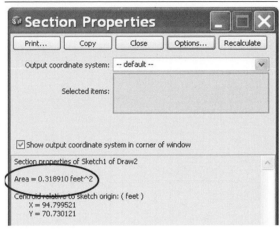

9.2.2 Calculating the Section Properties of a Shape

In mechanics of materials, it is often necessary to compute the moment of inertia of the cross-section of a structural member. For simple shapes, such as a rectangle or circle, easy formulas can be used for this computation. For compound shapes, such as a T-beam constructed from two rectangular shapes, the calculations are lengthier.

FIGURE 9.49

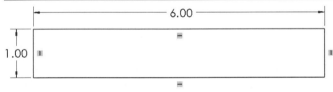

Open a new drawing. Choose A-Landscape paper without a sheet format displayed. Draw and dimension a rectangle, as shown in Figure 9.49.

FIGURE 9.50

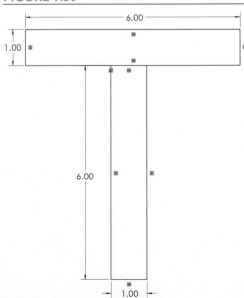

Add and dimension the second rectangle shown in Figure 9.50, making sure to snap the first point of the new rectangle to the edge of the first rectangle.

FIGURE 9.51

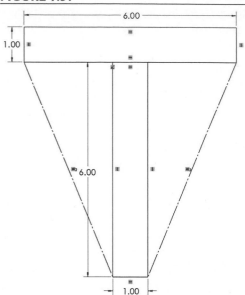

To center the second rectangle relative to the first, we could add a dimension. A better way is to add the centerlines shown in Figure 9.51, and add an Equal relation between them.

To compute the area or moments of inertia of the shape, it is not necessary to locate the drawing on the page. However, if we want to know the location of the centroidal axes, then we need to set a point at a known location.

Select the lower-left point of the second rectangle, as shown in Figure 9.52.

FIGURE 9.52

FIGURE 9.53

In the PropertyManager, set the coordinates of the point to x = 5 and y = 1, as shown in Figure 9.53.

In drawings, the lower-left corner of the page is the origin of the x-y coordinate system.

Select Tools: Section Properties from the main menu. A message will be displayed that the sketch has intersecting contours. Close the Section Properties dialog box.

The properties can be computed only for a sketch or drawing containing a single, closed contour.

Select the Trim Entities Tool from the Sketch group of the CommandManager. Trim away the overlapping portions of both rectangles, as shown in Figure 9.54.

FIGURE 9.54

Select Tools: Section Properties from the main menu.

FUTURE STUDY

Mechanics of Materials

The study of mechanics of materials involves determining *stresses, strains*, and *deformations* in bodies subjected to various loadings. Stress is the force per unit area acting on a point in a body, and the calculation of stress is necessary to predict failure of a structure. Strain is a measure of geometrical changes within a body. Stress and strain are related by properties of the material used. Deformations are related to strains in the body. While strain applies at a given point, deformation is the dimensional change of the entire body. For example, consider a 1 × 6-inch wooden plank resting on sawhorses near its ends. If the *span* of this beam is 6 feet, and a 175-pound man stands in the middle, how much will the plank move downward?

Using mechanics of materials concepts, it can be determined that the deflection of such a beam is:

$$\delta = \frac{PL^3}{48EI}$$

where:

    P = force = 175 lb

    L = span = 72 in

    E = *modulus* of the material. A typical value for wood = 1,500,000 lb/in^2

    I = moment of inertia of the cross-section

For a rectangular shape, the moment of inertia is:

$$I = \frac{1}{12} bh^3$$

So for our plank, with base b = 6 inches and height h = 1 inch, I = 0.50 in^4, and the deflection = 1.81 inches. If we could carefully place the plank so that it is resting on the 1-inch edge, then the values of b and h are switched. The moment of inertia increases to 18 in^4 and the deflection reduces to 0.05 inches. Therefore, the *bending stiffness* of the plank has increased by a factor of 36 by changing its orientation. The moment of inertia is increased by moving material to the greatest possible distance away from the *centroidal axis* (the axis passing through the centroid, or center of area, of the section). With wood construction, this is done by aligning the members appropriately, as with floor beams that are placed with their long dimensions perpendicular to the floor. In steel construction, wide-flange beams maximize bending stiffness by placing most of the material in the flanges, as far from the centroidal axis as possible.

When complex shapes are used, calculation of the moment of inertia can be cumbersome. First, the centroid must be located, and then the moments of inertia of simple regions of the shape must be calculated and adjusted for their distances away from the centroidal axis. Finally, the moments of inertia of the individual regions are summed.

The Section Properties Tool can be a useful tool for calculating and/or checking the value of moment of inertia.

FIGURE 9.55

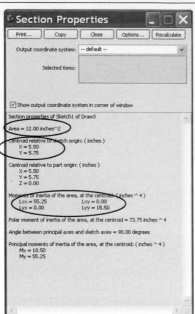

The section properties are displayed, as shown in **Figure 9.55**.

There are two principal moments of inertia calculated. The moment labeled "Lxx" is the moment of inertia about an axis through the centroid of the cross-section, parallel to the x axis. The location of this axis is given by the y coordinate of the centroid, 5.75 inches. Since a point at the bottom of the section was set at y = 1 inch, the axis is 4.75 inches above the bottom of the section, as shown in **Figure 9.56**. The moment of inertia about this axis is 55.25 in⁴. The other principal moment of inertia is about the axis shown in **Figure 9.57**. The value of the moment of inertia about this axis is 18.50 in⁴.

FIGURE 9.56

FIGURE 9.57

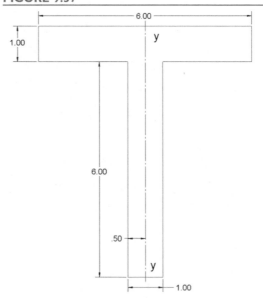

P9.1   Place the items from section 9.1 into the space shown here, which includes two 18-inch-square columns.

FIGURE P9.1

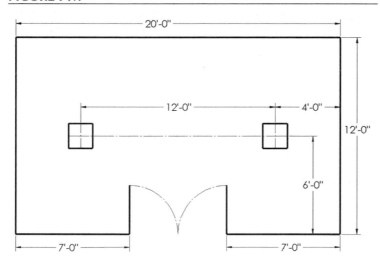

P9.2   Sketch a floor plan of your room or apartment, showing the locations of furniture.

P9.3   A builder desires to construct a house on the irregular lot shown here. The desired house will be rectangular in shape, 55 feet by 40 feet, with the front of the house 55 feet wide. The front of the house is to be parallel to the street. If local regulations require that there be 10 feet between property lines and any point on the house, can the proposed house be built on the lot?

FIGURE P9.3

Note: Three of the angular dimensions will be driven dimensions, since the length of all sides and one angle defines the lot shape. However, when property is surveyed, all lengths and angles are measured. Trigonometry is then used to ensure that the dimensions are consistent to within a certain tolerance. This method allows for incorrectly measured or recorded lengths and angles to be detected. The angles are measured in degrees and minutes (1 minute = 1/60 degree).

Hint: Use the Offset Entities Tool to offset the property lines 10 feet inward. You can then check to see if the house plan fits within the offset lines.

P9.4   Find the number of square feet in the lot described in P9.3.

(Answer: 6506 ft²)

P9.5   Find the location of the centroid and the principal moments of inertia of the channel shape shown here. Make a sketch showing the locations of the centroidal axes.

FIGURE P9.5

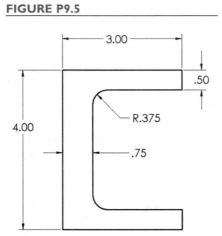

P9.6 Frames for industrial equipment are sometimes built out of prefabricated aluminum extrusions. These extrusions are cut to length and assembled with custom fasteners, providing engineers with the ability to construct custom framing without the need for complicated fixturing and welding. A typical cross-section for such an extrusion is shown in **Figure P9.6** (dimensions in inches). Create a drawing of this cross section, and use it to determine the cross-sectional area and principal moments of inertia of the shape.

FIGURE P9.6

P9.7   a) Fiber-reinforced composite plates are often made by stacking up layers of material with the fibers oriented in specified directions. Consider a 6-inch by 12-inch rectangular plate that is to be made up of two layers with fibers oriented at +30 degrees relative to the long axis of the plate, and two layers with fibers oriented at −30 degrees, as illustrated in **Figure P9.7A**. If the fiber-reinforced material comes on a roll that is 24 inches wide, with the fibers oriented along the length of the roll, determine the length L that is required for a rectangular portion of material from which the four required layers can be cut (see the example in **Figure P9.7B**). Since the material is expensive, try to place the layers in a manner that reduces the amount of scrap. Calculate the percentage of material that will be scrap.

FIGURE P9.7A

FIGURE P9.7B

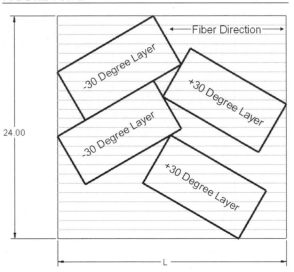

b) Suppose you are asked to evaluate a suggestion that the material be cut into rectangular sections from which two complete plates can be made (that is, four +30 layers and four −30 layers can be cut from the rectangular section). Can you develop a cutting pattern that will reduce the scrap percentage calculated in part a?

P9.8   A foam pad for an industrial product is shown in **Figure P9.8** (dimensions in inches). The pad will be mass-produced by die-cutting it from an 11-inch by 11-inch sheet of material. The die-cutting process requires that the parts must be separated by at least 0.1 inch from each other and the edge of the sheet. Design a layout for a die to cut as many parts as possible from the sheet, and use the information from the Section Properties Tool to determine the percentage of the raw material that will be scrap.

FIGURE P9.8

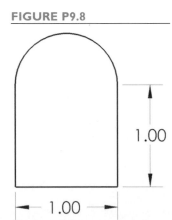

CHAPTER 10

Application of SolidWorks to Vector Mechanics

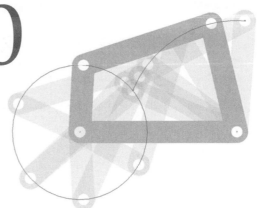

Introduction

Many engineering problems involve the manipulation of *vectors*. A vector is a representation of a quantity that is defined by both a magnitude and a direction, as opposed to a scalar quantity that can be defined by a magnitude only. For example, the speed of an object is a scalar quantity, while velocity is defined by both the speed and the direction of the object's motion, and therefore is a vector quantity. The SolidWorks 2-D drawing environment allows for easy graphical solution of vector problems.

10.1 Vector Addition

Consider the two forces acting on the hook as shown in **Figure 10.1**. We would like to find the resultant force, that is, a single force that affects the hook in an equivalent manner to the two forces. The resultant force is the vector sum of the two vectors **A** and **B**. (Vector quantities are usually denoted by bold type or by a bar or arrow over the symbol: **A** or $\overline{A}$ or $\vec{A}$.)

Chapter Objectives

In this chapter, you will:

- learn how to add vector quantities graphically,

- work with driving and driven dimensions,

- learn to solve for any two unknowns (magnitudes and/or directions) in a vector equation, and

- perform position analysis for some common mechanisms.

FIGURE 10.1

A = 60 lb

B = 70 lb

25°   60°

FIGURE 10.2

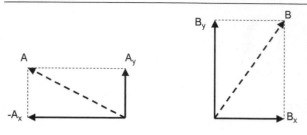

Analytically, the vectors can be added by adding the *components* of the two vectors. If we define an x-y coordinate system, then each vector can be broken into its x and y components, as shown in **Figure 10.2**.

The components of the resultant vector **R** are found by adding the components of **A** and **B**:

	x component	y component
A	$-60 \cos (25°) = -54.38$ lb.	$60 \sin (25°) = 25.36$ lb.
B	$70 \cos (60°) = 35.00$ lb.	$70 \sin (60°) = 60.62$ lb.
R	-19.38 lb.	85.98 lb.

From the components, the magnitude and direction of **R** can be determined (**see Figure 10.3**):

$$R = \sqrt{(19.38)^2 + (85.98)^2} = 88.14 \text{ lb.}$$

$$\theta = \tan^{-1}\left(\frac{85.98}{19.38}\right) = 77.3 \text{ deg}$$

FIGURE 10.3

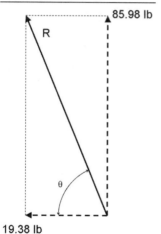

10.2   Vector Addition with SolidWorks

Open a new drawing. Choose A-Landscape as the paper size, with the "Display sheet format" box unchecked. Close the Model View Manager if it opens automatically.

FIGURE 10.4

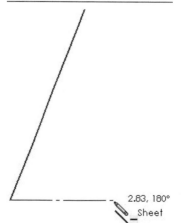

We must now choose scales: one for the magnitude of the vectors and another for the drawing. The vectors must be drawn to scale so that the length is proportional to its magnitude. If we choose to let 1 inch represent 10 pounds of force, then our two vectors will be 6 and 7 inches long. To fit onto our 8-1/2 × 11-inch drawing, it will probably be necessary to draw the vectors at one-half scale.

Right-click anywhere in the drawing area, and select Properties from the pop-up menu. Change the scale to 1:2.

(The default scale is defined in the template selected when creating the drawing.)

Select the Line Tool, and drag out a diagonal line to represent the first vector. From the starting point of the first vector, drag out a horizontal centerline, as shown in Figure 10.4.

Show the sketch relations by selecting View: Sketch Relations from the main menu.

Dimension the length of the vector line as 7 inches, and the angle between the vector line and the horizontal centerline as 60 degrees, as shown in Figure 10.5.

FIGURE 10.5

To add vectors graphically, we set them "tip-to-tail." That is, the second vector starts at the end of the first vector.

FIGURE 10.6

Draw the second vector line diagonally from the end of the first vector. (Be sure not to align this vector line with any of the dashed lines that appear. Doing so will add a constraint to the line that will have to be deleted before the vector's magnitude and direction can be defined.) Add another horizontal construction line, and add the 6-inch and 25-degree dimensions as shown in Figure 10.6.

The resultant vector extends from the starting point of the first vector to the ending point of the second vector.

Add the line representing the resultant vector, as shown in Figure 10.7, snapping to the endpoints. Add a length dimension to this line.

When you add this dimension, you will get a message that adding this dimension will make the drawing over defined, and asking if you want to make the dimension driven or leave it as driving (Figure 10.8). Click OK to make the dimension driven.

FIGURE 10.7

FIGURE 10.8

A driving dimension is one that helps control the size and position of the drawing or sketch entities, while a driven dimension does not.

Add a horizontal centerline and the angular dimension of the resultant vector, as shown in Figure 10.9. Click OK to make the dimension driving.

FIGURE 10.9

FIGURE 10.10

If the length of the resultant vector is not given to 3 decimal places, select that dimension and choose .123 from the pull-down menu in the PropertyManager, as shown in Figure 10.10. Also, set the angular dimension of the resultant vector to 2 decimal places.

The length of the resultant vector is 8.814 inches, corresponding to a magnitude of 88.14 pounds. This value and the angle of 77.3 degrees relative to the horizontal agree with the results found analytically.

10.3   Modifying the Vector Addition Drawing

The drawing just created is fine for calculating the magnitude and direction of the resultant vector, but visually is not clear. Vectors are usually shown with arrowheads to indicate their directions, and the line format of the vector lines can be changed to make them stand out from the construction and dimension lines. Also, the dimensions of the resultant vector can be shown differently to make them stand out from the input dimensions. We will modify the drawing to make it more understandable, and to allow easy modifications of the input quantities.

For this exercise we will use the Formatting and Line Format toolbars.

Right-click on any toolbar or the CommandManager. If they are not already displayed, click on Formatting and Line Format to display those toolbars.

FIGURE 10.11

We will first add arrowheads to the vectors. We will use a simple arrowhead: a single line segment.

Add a line to the end of the 7-inch vector line. Add a length dimension of 1 inch and angular dimension of 20 degrees, as shown in Figure 10.11.

Select both the vector line and the arrowhead line. From the Line Format toolbar, click on the Line Thickness Tool, as shown in Figure 10.12, and choose a thicker line style. If desired, change the color of the lines, using the Brush Tool to the left of the Line Thickness Tool. If you change the line color, clear the "Default" check box so that the line color that you specify overrides the default colors for under defined and fully defined entities.

FIGURE 10.12

These lines now stand out, as shown in Figure 10.13.

FIGURE 10.13

Hide the 1-inch and 20-degree dimensions, as shown in Figure 10.14.

Add and dimension arrowhead lines to the other vector lines. Hide the dimensions, and change the line thicknesses and colors, if desired, of the vector and arrowhead lines (Figure 10.15).

FIGURE 10.14

Use a different color for the resultant vector to make it stand out from the other vectors.

Modify the text of the dimensions by selecting them and changing the point sizes from the Formatting toolbar. Make the dimension text associated with the resultant vector larger than those of the other vectors. You can also make these text items bold by clicking the "B" Tool on the Formatting toolbar, as shown in Figure 10.16. To change the default color of the driven dimensions from grey to black, select Tools: Options from the main menu. Under the System Properties tab, select Colors, and change the color of Dimensions, Non Imported (Driven) to black (or to some other color, if desired).

FIGURE 10.15

FIGURE 10.16

FIGURE 10.17

The completed vector drawing is shown in **Figure 10.17**.

One of the main benefits of using the SolidWorks program to add vectors is that the vector drawing can be easily modified to solve different problems. We will use our drawing to add a different pair of vectors.

To make this drawing more useable, define the orientation of the vectors from a common reference: the +x direction.

Delete the 25- and 77.3-degree dimensions and associated reference lines. Add new construction lines and dimensions as shown in Figure 10.18. Add the 155-degree dimension first so that the 102.7-degree dimension is the driven one.

FIGURE 10.18

Dimensions can be changed from driven to driving or from driving to driven by right-clicking on the dimension and checking or un-checking Driven. Of course, if too many dimensions are driving, the drawing will be over defined.

Let's use this drawing to add the two vectors shown in **Figure 10.19**.

FIGURE 10.19

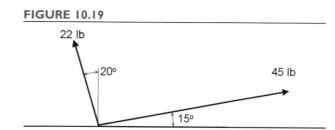

We will let 1 inch equal 1 pound for the vector's scale.

Change the dimensions of the input vectors to 22 inches and 110 degrees (from the horizontal) for the first vector, and 45 inches and 15 degrees for the second vector.

Of course, the drawing now extends beyond the edges of the sheet.

Right-click anywhere within the sheet borders, and select Properties. Set the scale to 1:5.

In the dialog box that appears, leave the check boxes as shown in Figure 10.20 and click OK.

FIGURE 10.20

This will allow the relative position of the dimension to stay the same, without changing the size of the text.

Drag the drawing into the sheet boundaries, and drag the dimensions into desired positions, as shown in Figure 10.21.

FIGURE 10.21

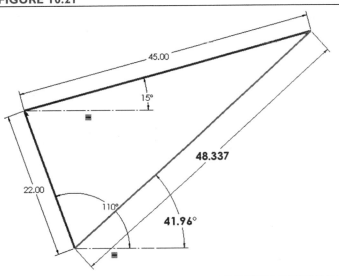

The resultant vector has a magnitude of 48.3 pounds, and is oriented at 42 degrees counterclockwise (CCW) from the x axis.

10.4 Further Solution of Vector Equations

In the previous example, two vectors were added to find the resultant vector's magnitude and direction. In a vector equation, any two unknowns (magnitudes and/or directions) can be determined. The following examples illustrate this concept.

A small plane can travel at an airspeed of 300 miles per hour. The flight path is to be at a heading of 15 degrees. (Heading is the angular direction measured CW from due North.) The wind is blowing from the WNW, as shown in Figure 10.22, at 60 mph. Find the plane's ground speed and the direction of the plane's travel relative to the air.

FIGURE 10.22

Wind

The vector equation for this problem is:

$$\mathbf{V}_{plane} = \mathbf{V}_{air} + \mathbf{V}_{plane/air}$$

where:

$\mathbf{V}_{plane}$ = absolute velocity of the plane (ground speed)—magnitude unknown and direction known

$\mathbf{V}_{air}$ = wind velocity—magnitude and direction known

$\mathbf{V}_{plane/air}$ = velocity of the plane relative to the air (airspeed)—magnitude known and direction unknown

Open a new A-size drawing and set the drawing scale to 1:50. We will let 1 inch equal 1 mile per hour.

FIGURE 10.23

Begin by drawing and dimensioning the vector representing the wind speed, as shown in Figure 10.23, near the bottom of the sheet.

It is helpful when we are dragging vectors together later to have a point fixed on the drawing.

Select the starting point of the first vector and click on the Fix icon in the PropertyManager.

If you need to move this point later, click on the point and the Fix relation will be listed in the PropertyManager. Select this relation and use the Delete key to remove it. The point can then be dragged to a new location.

FIGURE 10.24

Add the airspeed vector of 300 mph at an arbitrary angle, as shown in Figure 10.24.

Add the resultant (ground speed) vector at 15 degrees from vertical. The length is unknown, but make the line long enough so that the airspeed vector will intersect it when rotated into position, as shown in Figure 10.25. If desired, change the color of the resultant vector line to make it easy to distinguish from the airspeed and wind vectors.

Click and drag the endpoint of the airspeed vector until it intersects with the ground speed vector (see Figure 10.26).

With the Trim Entities Tool, trim away the end of the ground speed vector, as shown in Figure 10.27. (Make sure that the Trim option is set as "Trim to closest.")

FIGURE 10.25

FIGURE 10.26   **FIGURE 10.27**

Add dimensions for the ground speed magnitude and the direction of the airspeed, as shown in Figure 10.28.

FIGURE 10.28

The result is that the ground speed is 279 miles per hour. Relative to the wind, the plane must fly about 4 degrees east of due north to achieve the desired course.

Variations of this problem can be easily solved by changing the input quantities. For example, consider the case where the wind is blowing from due west at 70 mph.

Change the dimensions defining the wind speed vector, as shown in Figure 10.29.

Figure 10.29 shows the resulting vector equation. The ground speed is now 310 miles per hour, as the wind contributes to the east-to-west component of the plane's travel.

FIGURE 10.29

10.5 Kinematic Sketch of a Simple Mechanism

In Chapter 11, we will look at the application of the SolidWorks program to model mechanisms, with assemblies of 3-D component parts. Often, the first step in the design of a mechanism is the preparation of a *kinematic sketch*, a 2-D drawing showing simplified representations of the members. For example, a four-bar linkage, which is a common mechanism used in many machines, can be represented by four lines, as shown in **Figure 10.30**.

FIGURE 10.30

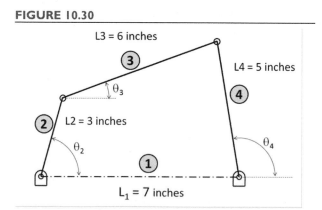

In a *kinematic analysis*, the velocities and accelerations of the links and points are calculated, based on the velocity and acceleration of the driving link. For example, if Link 2 is connected to a motor and rotated about its pivot point at a constant velocity, the angular velocities and accelerations of the other two moving links and the translational velocities and accelerations of points on those links can be calculated. The accelerations are important because force is proportional to acceleration. If the accelerations are known, the forces acting on the members and joints can be calculated. (Note: Although there are only three moving members, this mechanism is referred to as a four-bar linkage because it is connected to a fixed or ground link, which is usually called Link 1.)

The first step in any kinematic analysis is a *position analysis*. For a given position of the driving link (the angular position of Link 2), the positions of the other links must be calculated. Obviously, these positions can be calculated using trigonometry. For many mechanisms, a position analysis using trigonometry is surprisingly complex. A graphical solution is often utilized. The SolidWorks program is an excellent tool for graphical position analyses, in that dimensions can be easily changed and the effects on the rest of the linkage can be determined.

FIGURE 10.31

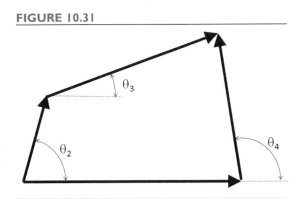

The links are often represented by vectors, as shown in **Figure 10.31**. In this case, all of the vector magnitudes (lengths) are known, as well as two of the vector angles (Link 1 is horizontal, and we will be performing the analysis for a given orientation of Link 2, the driving link). Therefore, we can solve the vector equation for the two unknown quantities, the angles θ_3 and θ_4.

In the previous sections, we used a SolidWorks drawing to create our sketches. For this example, we will sketch within a SolidWorks part file. For creating a quick sketch, the part file is often preferable because the sketches automatically scale. That is, since the sketch is not required to fit a paper size, we can enter the dimensions in any scale and simply change the viewing scale to display the entire sketch. (Sketches can also be created in assembly files. Often, a *layout sketch* within an assembly is used to size and/or position the components of a mechanism.) To print a copy of a sketch created in a part file, the screen capture utility can be used to copy and paste the sketch to a word processing file. The advantages of using a drawing file to create a sketch are that the drawing can be printed to the precise scale and with a title block, if desired, and the drawing can be formatted and annotated more easily.

Open a new part file. Select the Front Plane.

Beginning at the origin, draw three lines at arbitrary orientations, as shown in Figure 10.32, without adding vertical, horizontal, or perpendicular relations.

Add a centerline representing Link 1, and add a Horizontal relation to this line, as shown in Figure 10.33.

FIGURE 10.32

FIGURE 10.33

FIGURE 10.34

Add length dimensions, as shown in Figure 10.34. Be sure to add dimensions oriented with the links, not horizontal or vertical dimensions.

Add the angular dimension between Links 1 and 2, setting it to 90 degrees, as shown in Figure 10.35.

FIGURE 10.35

Add horizontal centerlines from which the angular positions of Links 3 and 4 will be measured, as shown in Figure 10.36.

FIGURE 10.36

Add an angular dimension between Link 3 and the adjacent horizontal centerline. A box will appear with the message that this dimension will over define the drawing. Select OK to make the dimension driven, as shown in Figure 10.37.

Add an angular driven dimension defining the position of Link 4. Set the decimal places of these dimensions to 2.

FIGURE 10.37

FIGURE 10.38

Select the Options Tool. Under System Options: Colors, scroll to Dimensions, Non Imported (Driven) and select Edit, as shown in Figure 10.38. Change the color to blue.

The completed sketch is shown in **Figure 10.39**.

When driven dimensions are shown in a different color, it makes clear that these dimensions do not control the position of the mechanism. Double-clicking on these driven dimensions will not allow their values to be edited.

While the angular dimensions found from **Figure 10.39** could be easily found with any 2-D CAD system, the advantages of the SolidWorks environment are seen when multiple variations of the mechanisms are to be found. For example, if we need to find θ_3 and θ_4 for a value of $\theta_2 = 45$ degrees, we need to change only that dimension.

FIGURE 10.39

FIGURE 10.40

Double-click on the 90-degree dimension and change its value to 45 degrees.

Notice that the driven dimensions θ_3 and θ_4 are updated accordingly, as shown in **Figure 10.40**.

A driving dimension can be changed to a driven dimension by right-clicking and selecting Properties, and then checking the Driven box, as shown in **Figure 10.41**.

Change the 45-degree dimension to Driven.

We can now click and drag the endpoint of Link 2, as shown in **Figure 10.42**, to investigate the range of possible motions. This mechanism is classified as a *crank-rocker* because one of the members that pivots about a fixed point can rotate 360 degrees (a crank), while the other member that pivots about a fixed point oscillates back and forth (a rocker). We notice that as the crank, Link 2, revolves, there are two positions for which Link 4, the rocker, is vertical. If we want to determine the positions of the crank for which this condition applies, then we need to make θ_4 a driving dimension.

FIGURE 10.41

FIGURE 10.42

Drag one of the links until Link 4 is almost vertical, as shown in Figure 10.43.

Right-click on the dimension defining the angular position of Link 4 and clear the Driven check mark. Change the dimension to 90 degrees.

FIGURE 10.43

We see in **Figure 10.44** that θ_2 equal to 59.95 degrees results in Link 4 being vertical.

When dragging Link 2 through its full range of motion, we found that there are two configurations for which Link 4 is vertical.

FIGURE 10.44

Set the 90° angle defining Link 4 to Driven. Drag the mechanism to approximately the position shown in Figure 10.45, and set the dimension back to Driving. Set the dimension to 90 degrees.

We see that $\theta_2 = 11.13$ degrees is another solution for which the rocker is vertical.

We can now examine the effect of changing the length of one of the links.

FIGURE 10.45

FIGURE 10.46

Set the mechanism back to its original position, with θ_2 as the driving dimension and θ_4 as a driven dimension, as shown in Figure 10.46.

Double-click the 6-inch dimension and change its value to 4 inches, as shown in Figure 10.47.

Change the 90-degree dimension to 150 degrees.

FIGURE 10.47

Note that the position of the links remains unchanged. This is because the position we have entered is not possible. The "No Solution Found" message in the status bar, as well as the color-coded highlights of the geometric conflicts, indicate this, as shown in Figure 10.48. To understand why the position is invalid, we can drag the links to determine their range of motion.

FIGURE 10.48

Set the 150-degree dimension to Driven, and drag the endpoint of Link 2 as far as possible in the counterclockwise direction, as shown in Figure 10.49.

FIGURE 10.49

We find that when θ_2 reaches about 123 degrees, Links 3 and 4 are aligned, preventing Link 2 from rotating further. This condition defines a *toggle position* of the mechanism. The mechanism is called a *double-rocker*, since neither link that pivots around a fixed point can rotate 360 degrees.

To find the precise location of the toggle position, a relation between Links 3 and 4 can be added.

Select Links 3 and 4, and add a collinear relation. The value of θ_2 shown in Figure 10.50, 123.20 degrees, defines the toggle position.

FIGURE 10.50

A second toggle position exists at $\theta_2 = -123.20$ degrees.

If you want to move the mechanism, it is necessary to either click on the collinear relation icon on the drawing and press the Delete key, select either Link 3 or Link 4 and delete the collinear relation from the PropertyManager, or use the Undo key to remove the relation.

PROBLEMS

P10.1–10.4 Find the vector **C**, which is the sum of vectors **A** and **B**, graphically. Check your results by adding the x and y components of the vectors. All angles are measured CCW from the +x axis, as shown in **Figure P10.1**.

FIGURE P10.1

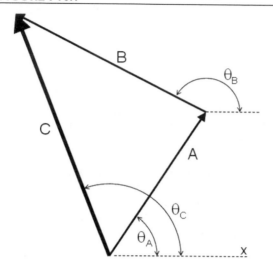

	Vector **A**		Vector **B**	
	Magnitude	Angle	Magnitude	Angle
p10.1	300 lb.	45°	150 lb.	90°
p10.2	10 m/s	0°	15 m/s	270°
p10.3	50 N	−10°	70 N	80°
p10.4	5 ft./s²	120°	4 ft./s²	300°

P10.5–10.8 Consider the vector equation **A** + **B** = **C**. Graphically find the unknown quantities for each set of vectors. All angles are measured CCW from the +x axis.

	Vector **A**		Vector **B**		Vector **C**	
	Magnitude	Angle	Magnitude	Angle	Magnitude	Angle
	A	θ_A	B	θ_B	C	θ_C
p10.5	200 lb.	25°	?	160°	300 lb.	?
p10.6	?	?	22 m/s	90°	36 m/s	78°
p10.7	?	130°	?	95°	50 N	110°
p10.8	8 ft./s²	112°	5 ft./s²	?	?	88°

P10.9 Consider two cars A and B approaching an intersection as shown in Figure P10.9. Using the vector equation $v_B = v_A + v_{B/A}$, find the velocity $v_{B/A}$, which is the velocity of B relative to A (that is, the apparent velocity of car B when viewed from car A). Comment on the result. Repeat the problem to find $v_{B/A}$ if car B is travelling away from the intersection at 40 mph.

FIGURE P10.9

40°

Car B:
v = 40 mph

Car A:
v = 30 mph

P10.10 The motion of a piston within a cylinder can be represented by the kinematic sketch shown in **Figure 10.10A**. Create this sketch, with the center of the rectangle representing the piston coincident with the vertical centerline. Examine the motion as the crank (the 2-inch long link) is rotated about the origin. Add the two angular dimensions shown in **Figure 10.10B**, with the connector angle (the 22.5-degree dimension) being a driven dimension. Find the extreme values of the connector angle.

FIGURE P10.10A   **FIGURE P10.10B**

P10.11 The mechanism illustrated in **Figure P10.11** is called an *offset slider-crank*.

As Link 2, the crank, rotates, Link 4, the slider, moves back and forth along a horizontal line. The distance L_4 is the horizontal distance from the pivot point of the crank to the pin joint between the connector, Link 3, and the slider. Link 1 is the ground, and the distance L_1 is the offset distance.

Construct a layout drawing of the mechanism, with $L_1 = 30$ mm, $L_2 = 50$ mm, and $L_3 = 150$ mm.

  a. Find L_4 and θ_3 for a crank angle $\theta_2 = 45$ degrees
      [Answers: 170.4 mm, 25.8 degrees)
  b. Find L_4 for values of θ_2 of 0, 90, 180, and 270 degrees.
  c. Find the minimum and maximum possible values of L_4, and the corresponding crank angles.

FIGURE P10.11

CHAPTER 11

Using SolidWorks in the Design and Analysis of Mechanisms

Introduction

A *mechanism* is an assemblage of *links* and *joints* that are connected together to achieve a desired motion task. The *links* provide the mechanical structure of the mechanism, while the *joints* provide the ability for the mechanism to move. Some typical kinds of mechanisms used by mechanical engineers are shown in **Figure 11.1**.

FIGURE 11.1

4-Bar Linkage

6-Bar Linkage

Slider-Crank

The ways in which we can connect links with joints and provide a mechanism with the ability to move are seemingly endless! Also, the links can take on any shape and size we desire, and the motion may change in complicated and seemingly unpredictable ways as we modify the links. The design of a mechanism may seem like a daunting task.

Chapter Objectives

In this chapter, you will:

■ learn some basic terminology associated with mechanism design,

■ use the assembly modeling techniques learned in Chapter 6 to develop a working model of a mechanism,

■ perform a motion simulation of a mechanism, and

■ export a motion simulation as a video file.

Computer-aided design (CAD) packages have become valuable tools in the design and analysis of mechanisms. The ability for the design engineer to adjust the size, shape, and interconnectivity of links and joints and quickly assess the impact on the mechanism has accelerated the design cycle of mechanisms. The SolidWorks program, with its ability to represent geometric constraints between structural components using assembly mates, is ideally suited to the design and virtual prototyping of complex mechanisms.

The remainder of this chapter will be devoted to the use of the SolidWorks program in the design of mechanisms through a case study involving the design of a *four-bar linkage*.

Approaching Mechanism Design with SolidWorks Assemblies

11.1

Consider the *four-bar linkage*, a classic mechanism used in engineering, shown in **Figure 11.2**.

It consists of three structural *links*, connected to each other and to fixed pivot points by *pin joints* that allow for rotating motion between the links. Though it only has three physical links, it is called a *four-bar linkage* because there is an implied fourth structural link (or *bar*) that connects the fixed ground points, as shown in **Figure 11.3**.

FIGURE 11.2

FIGURE 11.3

While the choice of a four-bar linkage as our preliminary design solution is an important step, the *parametric design problem* of selecting the appropriate link lengths and ground pivot locations to give us the desired motion is a difficult engineering task. We will develop a SolidWorks model to aid in this parametric design phase.

The features that we will employ are the *assembly* capabilities. The development of assembly models and the definition of assembly mates were covered in Chapter 6. Mated assemblies are an ideal tool for use in mechanism design, since the *joints* that provide the physical relationships between the *links* are analogous to the *mates* that define the geometric relationships between *parts*.

Think about two links connected by a pin joint, as shown in **Figure 11.4**.

FIGURE 11.4

The pin through the holes in the links allows for rotation between the links, giving a "scissors" action. The *mated assembly* representation of this type of link/joint assembly would involve two parts, with two mates serving the same geometric purpose as the pin joint in the physical linkage:

- The front face of one link is aligned with the back face of the other link using a *coincident mate*; this allows one link to "slide by" the other without any interferences between the surfaces.

- The hole in one link is aligned with the hole in the other link using a *concentric mate*; this keeps the holes aligned at their central axes.

By capturing the essential geometric constraints that underlie a *pin joint*, this sequence of two mates imparts the same motion capabilities to the solid model that are seen in the physical mechanism.

Using this simple two-link mechanism as a building block, complex mechanisms can be assembled and virtually tested. Mechanism motion can be tested and debugged without the need for physical prototypes to be constructed. The following section will step through the development of a model of a four-bar linkage.

11.2 Development of Part Models of Links

In this section, a step-by-step tutorial will lead you through the development of the part models required to make a "working" assembly model of a four-bar linkage. Four links, similar in shape but of different dimensions, will be constructed.

Open a new part, and select the Front Plane. Draw a rectangle centered about the origin, as shown in Figure 11.5.

Select the Tangent Arc Tool. Click on a corner point of the rectangle, and drag the arc away from the rectangle, as shown in Figure 11.6.

FIGURE 11.5

FIGURE 11.6

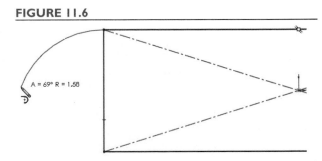

A = 69° R = 1.58

Drag the arc so that the endpoint snaps to the corresponding corner of the rectangle, as shown in Figure 11.7.

Draw a second tangent arc connecting the endpoints on the other end of the rectangle (Figure 11.8).

FIGURE 11.7   FIGURE 11.8

Use the Trim Entities Tool to remove the short sides of the rectangle, as shown in Figure 11.9. Add the dimensions shown in Figure 11.10.

FIGURE 11.9

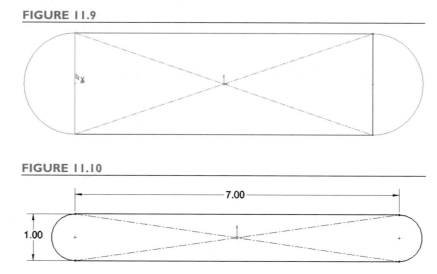

FIGURE 11.10

The sketch should be fully defined.

Extrude the sketch to a depth of 0.25 inches (Figure 11.11).

FIGURE 11.11

FIGURE 11.11

Select the front face of the part and choose the Circle Tool from the Sketch group of the CommandManager.

Two circles, concentric with the arcs at the ends of the link, will now be added. In order to activate the center points of the arcs as snap points, we must first "wake up" these features.

FIGURE 11.12

"Wake up" one of the center marks by momentarily holding the cursor on the circular edge of the part (Figure 11.12).

The center point of the arc will now be available as a snap point.

FIGURE 11.13

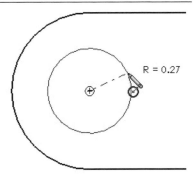

Sketch a circle centered at this snap point, as shown in Figure 11.13. Repeat this process at the other end of the link. Press the Esc key to turn off the Circle Tool. Select both circles, and add an Equal relation. Dimension one of the hole diameters as 0.50 inches, as shown in Figure 11.14.

The sketch should be fully defined.

Extrude a cut (type Through All) to create the holes.

FIGURE 11.14

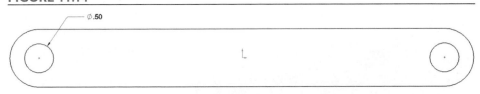

The finished part is shown in **Figure 11.15**.

FIGURE 11.15

Save the part as "Link1."

The easiest way to create each of the other three links is to modify the length of the first part and save it with a different name. All dimensions except the length should be identical for the four links.

Double-click Extrude1 in the FeatureManager. The dimensions defining this feature will be displayed, as shown in Figure 11.16.

FIGURE 11.16

FIGURE 11.17

Double-click the 7-inch length and change it to 3 inches. Rebuild the model (Figure 11.17).

Choose File: Save As from the main menu. Save this part as "Link2."

Create the other two links, using the lengths as shown in Table 11.1, and save them using the names "Link3" and "Link4." Close the part window.

TABLE 11.1 Lengths of the Four Link Parts

	Length
Link1	7 in.
Link2	3 in.
Link3	6 in.
Link4	5 in.

11.3 Development of the Assembly Model of the Four-Bar Linkage

The part models developed in Section 11.2 will now be used to construct an assembly model of the four-bar linkage. It is this assembly model that will allow us to perform parametric design and simple motion analysis of the mechanism.

FIGURE 11.18

Open a new assembly. If the PropertyManager shown in Figure 11.18 does not appear, open it by selecting the Insert Component Tool from the Assembly group of the CommandManager. Select Browse, and select the file Link1. Click the check mark to place this part.

This first part will be fixed in space—note the "(f)" beside its name in the Design Tree. We refer to it as the "ground link" since all motion of the other links will be relative to this fixed link.

Select the Insert Component Tool. Browse to select the file Link2. Click to place Link2 in the approximate position shown in Figure 11.19.

FIGURE 11.19

Note that you can click and drag Link2 to move it to a new position, as shown in Figure 11.20.

FIGURE 11.20

A coincident mate aligning the faces of these links will be defined.

Press Esc to cancel any selections made when moving Link2.

Select the Mate Tool to initiate a new mate.

The Mate PropertyManager will open.

Using the Rotate View Tool, rotate the parts so that the back faces of the links can be seen. Select the back face of Link2, as shown in Figure 11.21.

FIGURE 11.21

Switch to the Trimetric View, and select the front face of Link1, as shown in Figure 11.22.

FIGURE 11.22

Link2 will automatically move toward Link1 so that the selected faces show a coincident mate.

Click the check mark in the PropertyManager or the pop-up box to apply the mate.

A concentric mate between the holes must also be added to simulate the kinematic constraint of the pin joint.

Select the inside surfaces of the holes at the left end of the links (Figure 11.23).

A concentric mate will be previewed, as shown in Figure 11.24. Click the check mark to apply the mate. Click the check mark again to close the Mate PropertyManager.

FIGURE 11.23   **FIGURE 11.24**

The kinematic constraints of the pin joint are now fully defined. Link2 is still free to move, as long as the movements do not violate the mates that we have placed on it. We can experiment to see the type of motion still allowed under the constraints we have imposed.

Click and drag on Link2. Confirm that you can rotate the link though a full 360 degrees. Leave Link2 in the approximate position shown in Figure 11.25.

Note that the only unconstrained motion is rotation about the mated hole, as if the links were pinned together.

FIGURE 11.25

Select the Insert Component Tool. Select Link4 and place it into the assembly.

Select the Mate Tool.

In earlier chapters, we learned a shortcut method of selecting faces not visible from the current view orientation. We will use that technique to select the back face of Link4.

Move the cursor over Link4. Right-click, and pick Select Other, as shown in Figure 11.26.

FIGURE 11.26

The back face of the link is highlighted, as shown in Figure 11.27. Click the left mouse button to accept the selection. Select the front face of Link1. A coincident mate will be previewed; click the check mark to accept the mate.

FIGURE 11.27

Add a concentric mate to the inner faces of corresponding holes in Link1 and Link4. Close the Mate PropertyManager.

Click and drag Link4 to the approximate position shown in Figure 11.28.

FIGURE 11.28

Insert Link3 into the assembly. Add three mates:

1. **A coincident mate between the back face of Link3 and the front face of either Link2 or Link4,**

2. **A concentric mate between the inner faces of the corresponding holes of Link2 and Link3, and**

3. **A concentric mate between the inner faces of the corresponding holes of Link3 and Link4.**

Note that only one coincident mate is required. The coincident mate between the back face of Link3 and the front face of one of the other links completely defines the z-direction location of Link3. If another coincident mate is added between the back face of Link3 and the front face of another link, the assembly will be over-constrained and simulation problems may result.

FIGURE 11.29

The completed assembly is shown in **Figure 11.29**.

Click on Link2, and drag to rotate it. The link will rotate though a full 360°, and the motion of the other links can be observed. Save the assembly with the file name Linkage.

11.4 Creating Simulations and Animation with a Motion Study

In Section 11.3, we created a model of a four-bar linkage, and demonstrated the motion by manually dragging one of the links. In this section, we will learn to use the simulation tools built into the Assembly mode to create motion in our mechanism.

SolidWorks and its add-ins provide a number of levels of motion simulation. These include:

• Animation, available through the MotionManager, for animating the motion of assemblies with linear and rotary motors driving the motion,

- Basic Motion, available through the MotionManager, for studying the motion for assemblies with the effects of contact between members, gravity, and springs in addition to motors, and

- SolidWorks Motion, an add-in package, used for performing physical simulations with quantitative analysis of velocities, accelerations, and forces.

This section covers simple animation with the Animation tool; Section 11.5 will demonstrate more complex simulations using the Basic Motion tool. We will begin by adding motion to our linkage.

At the bottom left-hand corner of the assembly window, note that there are two tabs: a Model tab, and a Motion Study tab (labeled Motion Study 1). Note that if the Motion Study tab is not available, it can be added by selecting View:MotionManager from the main menu.

FIGURE 11.30

Click on the Motion Study 1 tab, as shown in Figure 11.30.

The MotionManager will appear, as shown in **Figure 11.31**. The MotionManager allows us to add simulation components such as motors, springs, and gravity to our model, lets us define the duration and resolution of our simulation, and gives us the ability to output animation files of our simulation.

The first step in our simulation will be to add a rotary motor to drive Link2.

FIGURE 11.31

Press the Esc key to deselect any selected part. Click on the Motor tool, as shown in Figure 11.32.

FIGURE 11.32

The Motor PropertyManager will open, as shown in **Figure 11.33**, with the
Component/Direction box highlighted.

FIGURE 11.33

Select the front face of Link2, as shown in Figure 11.34.

This adds a motor to the link, with the "drive shaft" of the motor oriented perpendicular to the selected face. The red arrow shows the direction of rotation;
this direction can be reversed by clicking the Reverse Direction button next to
the Component/Direction box if desired. The type and speed of the motion can
also be set; we will use the default Constant speed motion.

FIGURE 11.34

**Change the speed to 10 RPM, as shown in Figure 11.35. Click the check mark to close
the PropertyManager.**

FIGURE 11.35

We will adjust the simulation time to show us two complete revolutions of the motor.

Click and hold on the diamond-shaped key that marks the total simulation time, as shown in Figure 11.36. Drag it to the right, stopping at the 12-second mark (00:00:12), as shown in Figure 11:37.

FIGURE 11.36

FIGURE 11.37

Click on the Motion Study Properties tool, as shown in Figure 11.38.

FIGURE 11.38

The Motion Studies Properties Property-Manager will initiate. By default, the resolution of the simulation is set to 8 frames per second; this means that the simulation will produce a "snapshot" of the motion every 0.125 seconds (or every 7.5° of motor rotation, since our speed is 10 RPM). The higher we set the frames per second, the "smoother" the motion will appear in our simulation, but both more memory and more computation time will be required.

FIGURE 11.39

Set the Frames per second to 15, as shown in Figure 11.39, and click the check mark to close the PropetyManager. This will create a "snapshot" of motion at every 4° of motor rotation.

Save the assembly.

FIGURE 11.40

Click the Calculate tool to compute the simulation (Figure 11.40). The simulation will be computed, and displayed to the screen.

As long as no modifications are made to the model, the simulation need only be computed once; after this, it may simply be replayed, using the Play from Start, Play, and Stop controls, shown in Figure 11.41.

FIGURE 11.41

Click the Play from Start tool to replay the simulation.

Now that the simulation has been computed, an animation can be created from the simulation.

Click the Save Animation Tool, as shown in Figure 11.42.

FIGURE 11.42

The Save Animation to File dialog box will appear, as shown in **Figure 11.43**. This allows us to select the appropriate directory and filename for our video file. It also allows us to set some parameters that control file size and image quality, such as number of frames per second to be saved as we create our video file. For this application, we will accept the default value.

FIGURE 11.43

Browse to the appropriate file location, and click Save.

The Video Compression dialog box will appear, as shown in **Figure 11.44**. This dialog box allows us to select the image resolution of the video file. Higher-resolution images will result in larger file sizes. For this application, we will accept the default values for Compressor type and quality.

FIGURE 11.44

Click OK to close the Video Compression dialog box.

The simulation will be replayed again in the graphics window. While this is occurring, the video file will be written at the frames per second and compression level we specified.

As with the animation of the assembly explosion created in Chapter 7, the resulting file is an AVI file. It can be played with any standard video playback software. It is not associated in any way with the SolidWorks software or files; subsequent changes in your part or assembly files will not be reflected in the AVI file.

The animation file is essentially a "screen capture" of the Assembly window during the motion file; the motion will appear using the model orientation and scale shown in the Assembly window. In order to optimize the quality of the animation, the model should be oriented, centered, and scaled appropriately prior to creating the animation. If desired, the MotionManager can be collapsed prior to creating the animation file, in order to give the largest window area for the screen capture.

11.5   Investigating Mechanism Design

In this section, we will exploit the parametric modeling capabilities of SolidWorks to further investigate the design of mechanisms. In Section 11.4, we simulated the motion of the four-bar linkage using some given link lengths. In this section, we will modify the link lengths, and use motion simulation to predict the impact of the design changes on the motion of the mechanism.

The motion profile of a four-bar linkage is controlled by the relative lengths of the links. In the existing linkage, Link2 is capable of rotating through a full 360°. Such a link is called a "crank"; in a four-bar linkage a crank link will exist when the following equation is satisfied:

$$L + S < P + Q$$

where:

L = the length of the longest link
S = the length of the shortest link
P, Q = the length of the other two links.

In our linkage, L = 7 inches, S = 3 inches, P = 5 inches and Q = 6 inches. Since the condition is satisfied, at least one link (Link2) is a crank.

We will now modify our mechanism, and use a motion study to determine the impact of the redesign on the motion of the linkage.

FUTURE STUDY

Machine Dynamics and Machine Design

In this chapter, we have examined a four-bar linkage and performed some simple motion analysis. If we were designing this mechanism for an engineering application, many more questions would remain:

- How fast will the output link oscillate, if we know the speed of the input link?

- What size motor would be required to drive the input link?

- How will we transmit the rotational power of the motor to the input link?

Providing the answers to these questions would generally require the expertise of a mechanical engineer. The analysis of link speeds and accelerations is classified as a kinematics problem; extending this analysis to the sizing of motors is classified as a kinetics problem. Most engineering programs include basic courses in physics and dynamics, which address the essentials of kinematic and kinetic analysis. However, because of the great emphasis in mechanical

engineering on applying these principles to mechanisms and machinery, many mechanical engineering programs include upper-level courses in the advanced application of kinematic and kinetic analysis. Such a course is often called a *machine dynamics* course, and may include the study not only of linkages, but also of the dynamics of gears, cams, and other mechanical devices.

The question of power transmission requires further insight beyond kinematic and kinetic analysis. The choice of transmission also requires the investigation of various alternatives, such as a geared transmission, a system of belts and pulleys, or a chain and sprocket drive. The selection and sizing of the appropriate transmission system requires knowledge not only of kinematic and kinetic analysis, but also of the application of stress analysis to the transmission components. This type of analysis is often covered in mechanical engineering programs in *machine components* or *machine design* courses.

Click on the Model tab, as shown in Figure 11.45. Double-click on Link3 in the modeling window, change the length to 4 inches (Figure 11.46), and rebuild the model.

FIGURE 11.45

FIGURE 11.46

FIGURE 11.47

The modified model is shown in **Figure 11.47**.

Click on the Motion Study 1 tab to bring up the MotionManager. The Update Initial Animation dialog box will appear (Figure 11.48). Click Yes to close the box.

FIGURE 11.48

The model may temporarily appear to be disassembled, as in **Figure 11.49**. Disregard this.

FIGURE 11.49

In order to better simulate the physics behind the motion of the mechanism, we will use a higher-level motion simulation tool called Basic Motion.

FIGURE 11.50

Use the Type of Study pull-down menu to change the study type from Animation to Basic Motion (Figure 11.50). Click the Calculate tool to generate the motion simulation.

The linkage begins to move as before, until the Basic Motion simulation detects a position in which the linkage "locks up," as shown in **Figure 11.51**. Notice that in this position, Link3 and Link4 are aligned, and no further rotation of Link2 is possible. The position in which the mechanism "locks up" is called a *toggle position* (this was also demonstrated in Chapter 10). The presence of a toggle position is sometimes undesirable, since it prevents full rotation of the input link. Sometimes, however, toggle positions are designed into a linkage, so that it can be used as a clamp or fixturing device. Checking our equation with these new dimensions, we now have L = 7 inches, S = 3 inches, P = 5 inches, and Q = 4 inches. The equation is no longer satisfied, which is further evidence of the presence of a toggle position.

FIGURE 11.51

A second toggle position also exists in our mechanism. We will find it by further modifying our motion simulation.

FIGURE 11.52

Right-click on the RotaryMotor1 entry in the MotionManager tree, and select Edit Feature (Figure 11.52). Change the direction of motor rotation by clicking the Reverse Direction button (Figure 11.53), and click the check mark to close the PropertyManager. Click and hold on the diamond "key" associated with RotaryMotor1, as shown in Figure 11.54. Drag it to the beginning of the simulation, at time 00:00:00 (Figure 11.55).

FIGURE 11.53

FIGURE 11.54

FIGURE 11.55

Resetting the key allows the simulation to restart from the initial position, using the new motor direction for the motion.

Click the Calculate tool to compute the new simulation.

The mechanism will again stop at a toggle position, as shown in **Figure 11.56**.

FIGURE 11.56

We have examined the motion case of a four-bar linkage where $L + S < P + Q$ (where Link2 could rotate through $360°$), and also the case where $L + S > P + Q$ (where we found toggle positions). We will now explore the special case where $L + S = P + Q$.

Click on the Model tab to return to the assembly. Change the length of Link3 to 5 inches and rebuild.

The rebuilt model is shown in **Figure 11.57**.

FIGURE 11.57

Click the Motion Study 1 tab to return to the MotionManager. Click Yes to close the Update Initial Animation dialog box. Drag the diamond-shaped icon that marks total simulation time to the 24 second mark, as shown in Figure 11.58. If necessary, click the Zoom Out magnifying glass icon in the lower right corner of the screen to display the entire timeline. Click the Calculate tool to calculate the new simulation.

FIGURE 11.58

The motion of the linkage will appear unusual. If you watch it carefully, you will notice that the two complete motion cycles of Link2 produce two completely different motions for Link3 and Link4 (as shown in **Figures 11.59** and **11.60**). This mechanism is unstable. When the linkage passes through a certain point (when all links are aligned), there are two possible paths for the links to take.

FIGURE 11.59   **FIGURE 11.60**

In a stable four-bar linkage, the links can be assembled in either of two configurations, shown in **Figures 11.61** and **11.62**. Once assembled, the configuration is set. In an unstable linkage, both configurations are possible, and the mechanism may switch from one configuration to the other during the motion cycle. Unstable mechanisms are generally avoided in real motion applications, since their behavior can be unpredictable.

FIGURE 11.61   **FIGURE 11.62**

PROBLEMS

P11.1   A type of mechanism used in engineering systems is the *slider-crank* (see **Figure P11.1A**). In the slider-crank, the input link (crank) rotates continuously through a full 360 degrees, while the output slider slides along a fixed surface. Among other things, the *slider-crank* is the working schematic for a single cylinder of an internal combustion engine. Create a working SolidWorks assembly model of this mechanism, using the dimensions shown in **Figure P11.1B**. The dimensions are mm. (Hint: The bottom of the slider is always coincident with the ground plane.)

FIGURE P11.1A

FIGURE P11.1B

P11.2   The mechanism described in P11.1 can be modified by offsetting the pivot location of the crank from the path of the slider, creating an *offset slider-crank*, as shown in **Figure P11.2A**. This type of mechanism is sometimes called a *quick-return* mechanism, since the back-and-forth motion of the slider is different in one direction from the other. Create a working model of this mechanism, using the dimensions shown in **Figure P11.2B**. The dimensions are mm.

FIGURE P11.2A

FIGURE P11.2B

P11.3 Additional links can be added to a four-bar linkage to modify the motion of the links. A common engineering example is the *six-bar linkage*. **Figure P11.3A** shows a model of a six-bar linkage that simulates the operation of a car's windshield wipers. The Crank is attached to the wiper motor, which rotates at a constant speed. The lengths of the Crank and Connector 1, along with the attachment location to the blade, are selected so that the rotational motion of the wiper motor is converted to the desired oscillating motion of the first wiper blade. Connector 2 links the two blades together so that they move parallel to each other.

FIGURE P11.3A

Model the components shown in **Figure P11.3B**, and assemble them into the mechanism, using the spacing of ground joints shown in **Figure P11.3C**.

FIGURE P11.3B

FIGURE P11.3C

P11.4   In the 1700s, noted engineer James Watt devised and patented a mechanism for generating straight-line motion with a rotational input. In this mechanism, shown in **Figure P11.4A**, the center point of Link 3 will trace out a straight line in space as the input link (Link 2) rotates, as illustrated in **Figure P11.4B**. Originally designed for guiding the stroke in a steam engine piston, this mechanism is currently used to guide axle motion in automotive suspension applications[1]. Develop a working model of the *Watt Straight Line Mechanism*. Use a length of 8 inches for Links 2 and 4, and a length of 4 inches for Link 3. The distance between the fixed pivot points is 16 inches. (Note: This is considered a *double rocker mechanism*; the input link is *not* able to rotate through a full 360 degrees.)

FIGURE P11.4A

FIGURE P11.4B

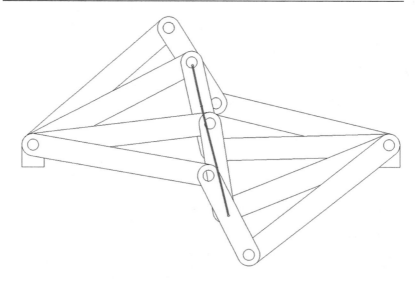

---

[1] Robert L. Norton, *Design of Machinery,* 2nd ed. (New York: McGraw-Hill, 1999).

P11.5 A *geneva mechanism*, which is illustrated in **Figure P11.5A**, is used to transform constant rotational motion into intermittent motion. Among other uses, this type of mechanism is used to control the motion of an *indexing table* in an assembly line. An indexing table will remain stationary for a period of time, and then rotate a fixed amount.

Model the components necessary to create a model of the geneva mechanism:

FIGURE P11.5A

a. The *geneva wheel*, as detailed in **Figure P11.5B** (thickness = 0.25 inches)

FIGURE P11.5B

b. The *locking disk*, as detailed in **Figure PII.5C** (thickness = 0.25 inches)

FIGURE PII.5C

c. The *crank*, as detailed in **Figure PII.5D** (thickness = 0.25 inches)

FIGURE PII.5D

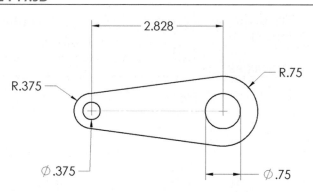

d. The *ground link*, as shown in **Figure PII.5E** (thickness = 0.25 inches)

FIGURE PII.5E

e. A cylindrical *shaft*, 0.75 inches in diameter by 2 inches long

f.   A cylindrical pin, 0.375 inches in diameter by 0.50 inches long.

Create the subassembly shown in **Figure P11.5F** from a shaft and the geneva wheel. Add mates so that the wheel is completely fixed relative to the shaft.

FIGURE P11.5F

Create the subassembly shown in **Figure P11.5G** from a shaft, the crank, the pin, and the locking disk. Add mates so that the crank, pin, and disk are all completely fixed relative to the shaft.

Create a new assembly with the ground link as the fixed component. Add the two subassemblies created above, and create mates so the shafts can rotate within the ground link. Orient the assembly so that the pin on the crank subassembly is aligned with one of the slots on the geneva wheel.

FIGURE P11.5G

To view the motion, click on the Move Component tool (**Figure P11.5H**). In the PropertyManager, click the Physical Dynamics option, and move the Sensitivity slider to the right, approximately 3/4 of the way toward Max (**Figure P11.5I**). Drag the crank in a circle to view the motion, as shown in **Figure P11.5J**.

FIGURE P11.5H   **FIGURE P11.5I**   **FIGURE P11.5J**

P11.6 Develop a motion simulation of the geneva mechanism model created in P11.5. In order to do this, perform the following steps:

1. Create a motion study, and add a rotary motor to the crank subassembly.

2. Use the pull-down menu to change the motion type to Basic Motion.

3. Add a 3-D Contact between the pin and the geneva wheel by:

   a. Selecting the Contact tool from the MotionManager (**Figure P11.6A**)

   FIGURE P11.6A

   b. Clicking on the pin (from the crank subassembly) and the geneva wheel part to define a contact between these components in the PropertyManager (**Figure P11.6B**)

   FIGURE P11.6B

   Select the pin

   Select the geneva wheel

   c. Clicking the check mark to close the 3-D Contact PropertyManager

The motion can now be simulated. Set the properties such that four full rotations of the crank are simulated. Export an AVI file of the motion.

P11.7 A Scotch yoke mechanism is used to convert continuous rotary motion into reciprocating linear motion. The mechanism is shown in **Figure P11.7A**. Create a simulation of the Scotch yoke mechanism by performing the following steps.

FIGURE P11.7A

a.   Using the crank, shaft, and pin parts from Problem P11.5, create the subassembly shown in **Figure P11.7B**. All parts should be fully constrained in the assembly. Save the assembly file as *scotchcrank*.

FIGURE P11.7B

b. Sketch the slotted T-shaped part, as shown in **Figure P11.7C**. Extrude it to a height of 0.25 inches, and save the part file as *yoke*.

FIGURE P11.7C

c. Create the base part shown in **Figure P.11.7D**. Save it with the name *scotchbase*.

FIGURE P11.7D

d. Assemble the Scotch yoke mechanism to allow for motion. (Hint: Use a tangent mate between the pin in the *scotchcrank* subassembly and the slot in the *yoke* part, and use coincident mates to establish the sliding motion between the slot in the *scotchbase* part and the *yoke* part.)

e. Add a 10-rpm motor to the crank, and create an animation of the motion showing two complete rotations of the crank. Save the animation as an AVI file.

CHAPTER 12

The Use of SolidWorks as a Tool for Manufacturing: Mold Design and Sheet Metal Parts

Introduction

A design engineer must always consider the method of manufacture when designing any part. Failure to do so may result in part designs that are more expensive to make than necessary, have high scrap rates, or cannot be made at all.

Some manufacturing processes create unique challenges from a solid modeling standpoint. For example, when designing a mold, the shape of the part to be molded must be removed from the interior of the mold, usually with the dimensions adjusted to allow for shrinkage of the part during the cool-down portion of the molding cycle. Sheet metal parts are cut from flat material, and then bent into the final shape. Therefore, the part definition must include both the flat shape and the finished geometry. The SolidWorks program has specialized tools for working with molds and sheet metal parts.

Chapter Objectives

In this chapter, you will:

- create a cavity within a mold base,

- create and modify two mold halves that are linked to the mold base with a cavity,

- make a simple sheet metal part,

- learn how to show a sheet metal part in either the flat or bent state, and

- make a drawing of a sheet metal flat pattern.

12.1 A Simple Two-Part Mold

In this exercise, we will make a mold to produce a simple cylindrical part. We will begin by making the part itself.

Open a new part. Draw and dimension a 3-inch diameter circle in the Right Plane, centered at the origin. Extrude the circle using a midplane extrusion, as shown in Figure 12.1. Set the total depth of extrusion to 5 inches. Save this part with the file name "Cylinder." Do not close the part window.

The next step is to create a *mold base,* from which a cavity in the shape of the part will be removed. The mold base must be large enough to completely enclose the part.

Open a new part. Draw and dimension a 4-inch by 6-inch rectangle in the Top Plane centered about the origin (Figure 12.2).

FIGURE 12.1

FIGURE 12.2

Extrude the rectangle with a midplane extrusion, as shown in Figure 12.3. Set the total depth of extrusion to be 4 inches. Save this part with the file name "Base." Do not close the part window.

We will now place the part within the mold base.

From the main menu, select File: Make Assembly from Part, as shown in Figure 12.4. Click OK to accept the default assembly template.

FIGURE 12.3

FIGURE 12.4

A new assembly window will open, and the base will appear as the first component to be inserted. (Note: If the base part is not previewed in the graphics window, check the "Graphics preview" box under the Options tab of the PropertyManager.)

If the origin is not visible in the graphics area, select VIew: Origins from the main menu to display it.

Click the check mark to place the base into the assembly, with the origin of the part placed at the origin of the assembly.

Note that placing the origin of the first component of an assembly at the origin of the assembly itself is not required, since the first component will be fixed and subsequent parts will be located relative to the first part. However, placing the first part in this manner is good practice, since the Front, Top, and Right Planes of the assembly, as well as the origin, can be used as mating entities.

FIGURE 12.5

Select the Insert Components Tool. Select the cylinder part from the list of open files, as shown in Figure 12.5.

Click the check mark to place the cylinder into the assembly, with the origin of the cylinder placed at the origin of the assembly.

The cylinder is now centered within the base, as shown in the wireframe view of **Figure 12.6**. Rather than work in a wireframe mode, it is helpful to display the base as transparent.

Save this assembly with a file name of "Mold Assembly."

FIGURE 12.6

FIGURE 12.7

Right-click on the part name "Base" in the FeatureManager. Select Appearance: Appearance. In the PropertyManager, select the transparency slider bar, as shown in Figure 12.7. Move this bar to the right to display the base in varying degrees of Transparency. Click the check mark to apply the desired transparency.

The cylinder can now be seen within the base, as shown in Figure 12.8.

FIGURE 12.8

We will now create the cavity. Since the cavity will be created in the base, the base must be selected for editing.

Select the Base from the FeatureManager. Choose the Edit Component Tool from the Assembly group of the CommandManager, as shown in Figure 12.9.

From the main menu, select Insert: Molds: Cavity, as shown in Figure 12.10.

In the PropertyManager, select the cylinder part from the FeatureManager as the component defining the cavity. Set the "Scale about" option to "Component Centroids" and the scale to 3%, as shown in Figure 12.11. Click the check mark to create the cavity. Click on the Edit Component Tool to end the editing of the base.

FIGURE 12.9

FIGURE 12.10

FIGURE 12.11

The scale factor causes the cavity to be larger than the finished part. Most molding materials shrink during cure or cooling, so the scale factor allows for that shrinkage.

The cavity can now be seen within the base, as shown in **Figure 12.12**. If you look closely, you can see that there are gaps between the edges of the part and the corresponding edges of the cavity, because of the shrink factor.

FIGURE 12.12

The base, which now includes the cavity, needs to be split into two halves. For this simple mold, the two halves will be identical, and so we could cut away half of the original base part. However, if the two mold halves will be different (as in the next exercise), then copies of the base must be made. This procedure of creating *derived parts* is illustrated here.

Select the Base from the FeatureManager. From the main menu, select File: Derive Component Part, as shown in Figure 12.13. In the New SolidWorks Document dialog box that appears, click OK to select the part template.

FIGURE 12.13

A new part window is opened, and a copy of the base (including the cavity), is created. The advantage of creating a derived part rather than simply saving a copy of the base part is that associativity is maintained. That is, if a change is made to the cylinder part, then the cavity in the assembly, the base, and the derived mold half part are all updated.

Open a sketch on the front face of the new part. Select the Line Tool, and draw a line completely through the part, passing through the origin, as shown in Figure 12.14. Extrude a cut with a type of Through All, with the direction to cut arrow pointing up.

The resulting part is shown in **Figure 12.15**.

FIGURE 12.14

FIGURE 12.15

FIGURE 12.16

Save the part with a file name of "Mold Half."

To illustrate the associativity of the parts, open the cylinder file and change the diameter of the part from 3 to 2 inches. Rebuild the part. If you switch immediately to the mold half part, then no changes will be seen. However, if you switch to the mold assembly, then the cavity will be updated. Then, switching to the mold half will cause the change to be reflected, as shown in **Figure 12.16**.

Two of these mold halves can now be assembled, and mold-level features (fill and vent ports, alignment pins, etc.) can be added.

12.2   A Core-and-Cavity Mold

In this exercise, a two-piece mold for making the card holder from Chapter 5, shown in **Figure 12.17**, will be created. The shape of this part requires that the mold geometry consist of a *core* half (**Figure 12.18**), with features protruding outward from the parting line, and a *cavity* half (**Figure 12.19**), with features cut inward from the parting line.

FIGURE 12.17

FIGURE 12.18

FIGURE 12.19

Our procedure will be similar to the one followed in the previous section. Since there are several files involved in this procedure, it is helpful to consider the process steps, which are illustrated in **Figure 12.20**. An interim assembly will be made from the mold base and part, so that the cavity can be placed in the mold base. From that assembly, copies of the mold base (with the cavity) will be derived. These copies will be modified to become the two mold halves. Finally, the two mold halves will be brought together into an assembly.

FIGURE 12.20

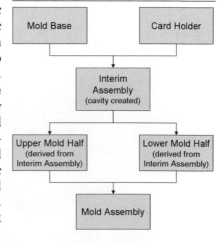

The first step of the process will be to create the mold base, which must be sized so that the card holder part fits completely within its boundaries.

Open a new part. In the Top Plane, draw and dimension a rectangle centered about the origin, as shown in Figure 12.21.

FIGURE 12.21

The dimensions shown are each 1 inch greater than the corresponding dimensions of the card holder, allowing for 1/2 inch clearance between the part and the mold edges.

The height of the mold base will be selected to allow for 1/4 inch above and below the part.

Extrude the rectangle upward a distance of 1.5 inches, as shown in Figure 12.22.

Choose Edit: Appearance: Appearance from the main menu, and move the Transparency slider bar toward the right, as shown in Figure 12.23.

FIGURE 12.22   FIGURE 12.23

The transparent part is shown in Figure 12.24.

Save this part as "Mold Base."

FIGURE 12.24

Select File: Make Assembly from Part from the main menu to open a new assembly. Insert the mold base into the assembly by clicking the check mark. Choose the Insert Components Tool and browse to locate the card holder part. Click to place it in the assembly in the approximate position shown in Figure 12.25.

We will now align the card holder with the mold base.

Rotate the view so that the bottom surfaces of the card holder and the mold base are visible, as shown in Figure 12.26.

FIGURE 12.25

FIGURE 12.26

FIGURE 12.27

Choose the Mate Tool. Select the bottom surface of the card holder (Figure 12.27), and then the bottom surface of the mold base. A coincident relation will be previewed. Select the Distance Mate icon from the PropertyManager (Figure 12.28). Set the offset distance between the two surfaces to be 0.25 inches. If necessary, check the Flip Dimension box so that the bottom surface of the card holder is above the bottom surface of the mold base, as previewed in Figure 12.29. Click the check mark to apply the mate.

FIGURE 12.28

FIGURE 12.29

Select the Front Planes of the card holder and mold base, as shown in Figure 12.30. Add a coincident mate, as shown in Figure 12.31.

Add a coincident mate between the Right Planes of the card holder and mold base. Close the Mate PropertyManager.

The card holder is now placed within the mold base, as shown in Figure 12.32.

FIGURE 12.30

FIGURE 12.31

FIGURE 12.32

Save the assembly as "InterimMoldAssy."

Select the mold base from the FeatureManager. Choose the Edit Component Tool from the Assembly group of the CommandManager.

From the main menu, select Insert: Molds: Cavity. In the PropertyManager, select the card holder as the component defining the cavity. Set the "Scale about" option to Component Centroids and the scale to 1%, as shown in Figure 12.33. Click the check mark to create the cavity. Click the Edit Component Tool to end the editing of the mold base.

FIGURE 12.33

Select the mold base from the FeatureManager. From the main menu, select File: Derive Component Part. Click OK in the New SolidWorks Document dialog box to select the part template.

A new part window is opened, and a copy of the mold base is created.

Display the part in the Wireframe mode, as shown in Figure 12.34.

FIGURE 12.34

We will now cut away the top portion of the mold to create the lower mold half. This cannot be done with a simple extruded cut, since that would also cut through the core feature that will form the underside of the card holder. Rather, we will perform two separate cutting operations to achieve the desired geometry.

Select the top surface of the part and choose the Sketch Tool from the Sketch group of the CommandManager to open a sketch. Select the four edges of the cavity shown in Figure 12.35. (Remember to use the Ctrl key to select multiple entities.)

FIGURE 12.35

Select the Convert Entities Tool, which will create lines from the selected edges.

Extrude a cut with a type of "Up to Next." This will cut away the top of the mold, but only to the cavity, as shown in Figure 12.36.

In order to cut away the material around the edges of the mold half, it is necessary to use a different type of cut: one that cuts down to the parting surface.

FIGURE 12.36

Open a new sketch on the top surface. Convert the eight edges shown in Figure 12.37 into lines, using the Convert Entities Tool.

Extrude a cut, with a type of "Up to Surface." For the surface, choose the bottom surface of the cavity, which corresponds to the parting line, as shown in Figure 12.38.

The completed mold half is shown in Figure 12.39.

FIGURE 12.37

FIGURE 12.38

FIGURE 12.39

Save this part as "MoldHalfLower."

We will now create the upper mold half.

Switch back to the assembly containing the mold base and the card holder (InterimMoldAssy). Select the mold base, and from the main menu, select File: Derive Component Part. Click OK to select the Part template.

In the new part, select the Section View Tool from the Heads-Up View Toolbar. Click the check mark to accept the Front Plane as the section plane. Zoom in to the bottom of the cavity, and click on the bottom surface to select it, as shown in Figure 12.40. Open a sketch on this surface by selecting the Sketch tool from the Sketch group of the CommandManager.

Turn off the Section View and select the Top View. Using the Corner Rectangle Tool, drag out a rectangle such that the entire part is enclosed within the rectangle, as shown in Figure 12.41.

FIGURE 12.40

FIGURE 12.41

FIGURE 12.42

Select the Extruded Cut Tool, and set the type of cut to Through All. Make sure that the direction of the cut is downward. Rotate the view so that the bottom of the part is visible, as shown in Figure 12.42, and click the check mark.

Since this cut will produce two separate solid bodies (the mold half and part of the core), you are prompted to identify which of the bodies you want to keep. In our case, we do not want to keep the core with the upper mold half. In the dialog box, checking the Selected bodies box allows you to choose which of the solid bodies you want to keep, and which will be deleted.

Choose the Selected bodies option. Check the box for Body 1 (the mold half) and clear the box for Body 2 (the core), as shown in Figure 12.43. The selected body is highlighted. Click OK to complete the cut.

The resulting part is shown in Figure 12.44.

Save this part as "MoldHalfUpper."

Open a new assembly. Insert the two mold halves, as shown in Figure 12.45. Make sure to place the first mold half at the origin of the assembly.

Add three mates to align the mold halves together, as shown in Figure 12.46.

FIGURE 12.43

FIGURE 12.44

FIGURE 12.45

FIGURE 12.46

Choose the Section View Tool from the Heads-Up View toolbar.

FUTURE STUDY

Materials and Processes

Our simple two-piece mold includes a cavity the shape of the finished part. If we created this mold for making a few prototype parts from a room-temperature curing material, such as polyurethane, then we could add holes for bolting the halves together, a hole for pouring in the material, and a vent hole, and our mold would be functional.

Most molds require many other features, however, and tooling design is an important function at any manufacturing company. If a plastic part is to be injection-molded, then the injection points and vent locations must be carefully designed so that the molten plastic fills the cavity completely. The plastic's melting and cooling temperatures and its resistance to flow must be considered when designing both the part and the mold. The tolerances required for the finished part might require that a filler be added to the material for dimensional stability. Ejector pins might need to be added to help remove the part from the mold. (Note that in our example, even though the part walls are tapered, the shrinkage of the part onto

the core will result in forces that will need to be overcome in order to remove the part from the lower mold half.) Since the plastic must be injected hot but allowed to cool before removal from the mold, cooling lines for circulating water are usually added to the mold halves.

Other materials have different processing requirements. Composite materials used for automotive materials are mostly compression molded, in which the raw material is placed between two mold halves and formed by applying pressure with a hydraulic press. Lower-quantity parts can be produced by resin transfer molding, in which dry fabric is placed in a mold and liquid resin is pumped in under low pressure.

Often the choice of a process depends on the quantity of parts to be made. A high-quality tool for injection molding can cost tens of thousands of dollars, but if this cost can be spread over 100,000 or 1,000,000 parts, then the fast cycle times resulting from a good mold design can result in significant cost savings.

Click the check mark to accept the Front Plane as the section plane, as shown in Figure 12.47.

The mold cavity can now be clearly seen.

Save the file with the name "Card Holder Mold Assembly."

Suppose that we now want to change the thickness of the molded part. Often, if a material change is specified, the thickness will need to be changed, since the flow of the material in the mold is a limiting factor on the thickness.

FIGURE 12.47

FIGURE 12.48

Open the card holder part. Right-click on the shell feature in the FeatureManager. Select Edit Feature, and double the thickness, from 0.06 to 0.12 inches. Rebuild the model.

The modified part is shown in **Figure 12.48**.

Open, or switch to, the interim mold assembly, then the two mold halves, and finally the mold assembly. At every step, the models will be updated to reflect the new thickness.

The updated mold assembly is shown in **Figure 12.49**. Of course, the mold we created is not usable without some way to get material into the mold. The manner in which this is done depends on the molding process, as discussed in the Future Study box. Many other features may also be required for the mold to be usable. Mold design is a very specialized field, combining mechanical design with material science. However, an essential part of any mold design is the creation of the mold cavity and the separation of the core and cavity mold parts, such as we have done in these exercises.

FIGURE 12.49

12.3 A Sheet Metal Part

In this exercise we will create the sheet metal part shown in **Figure 12.50**. The part can be shown in either the bent or flat state.

FIGURE 12.50

FIGURE 12.51

Open a new part. From the Options Tool, select Document Properties: Units. Set the unit system to MMGS (millimeters), and the number of decimal places to None for both the length and angular values.

Select the Right Plane. Sketch and dimension the three lines shown in Figure 12.51, placing the origin at the midpoint of the 70 mm vertical line. **Use the vertical centerline to align the endpoints of the diagonal lines.**

Note that we are not adding radii to the sharp corners. In sheet metal parts, *bends* are added as separate features to a part.

Select the Extruded Boss/Base Tool from the Features group of the CommandManager. Set the type to Mid Plane, and the extrusion depth to 100 mm. Since the sketch is open, a thin-feature extrusion will be created. Set the thickness to 2 mm, as shown in Figure 12.52.

FIGURE 12.52

Change the direction of the thickness if necessary so that the dimensions apply to the outside of the part, as shown in Figure 12.53. Click the check mark to complete the extrusion.

FIGURE 12.53

There are a number of tools that are specific to sheet metal parts. We can access these tools through the Sheet Metal toolbar or by adding the Sheet Metal tools to the CommandManager.

Right-click on one of the CommandManager tabs, as shown in Figure 12.54. In the list of tool groups that is displayed, click on Sheet Metal, as shown in Figure 12.55.

The Sheet Metal tools are now included in the CommandManager.

FIGURE 12.54

FIGURE 12.55

Select the Sheet Metal tab. Select the Insert Bends Tool, as shown in Figure 12.56.

FIGURE 12.56

FIGURE 12.57

FIGURE 12.58

Select the middle face as the one that will remain flat during the bending, as shown in Figure 12.57. Set the radius to 4 mm, and set the K-factor to 0.5. Leave the auto relief checked with the type as Rectangular and the factor set to 0.5, as shown in Figure 12.58. Click the check mark to complete the operation.

The resulting bent geometry is shown in **Figure 12.59**.

FIGURE 12.59

The K-factor is used in converting the bent geometry into the flat geometry. A K-factor of 0.5 means that the length of the flattened metal will be calculated based on the arc length at the mid-thickness of the metal. This is typical for relatively large bend radii. If extremely tight bends are desired, then the K-factor may need to be adjusted to get a better correlation between the flat geometry and bent geometry dimensions.

The FeatureManager now shows several new items, as shown in **Figure 12.60**. These new items include Sheet-Metal, where the bend radii, K-factor, and Auto Relief factors are defined, Flatten-Bends, where individual bends can be edited, and Process-Bends, which restores the bends when the part is rebuilt. The Flat-Pattern, which is shown as suppressed, contains the information to show the flattened configuration of the part.

The part can now be toggled between the flat and bent configurations using the Flatten Tool.

Click on the Flatten Tool (Figure 12.61) to show the part in the flattened state, as shown in Figure 12.62 (shown in Wireframe to show the bend lines clearly).

FIGURE 12.60

FIGURE 12.61

FIGURE 12.62

Notice that the Flat-Pattern is now active in the FeatureManager. Expanding the Flat-Pattern shows the bend lines and individual bend properties, as shown in **Figure 12.63**.

FIGURE 12.63

Click the Flatten Tool again to display the part in the bent state.

FIGURE 12.64

With the part in the bent configuration, the bends themselves can be suppressed with the No Bends Tool, as shown in **Figure 12.64**. This tool allows you to toggle between the geometry with sharp corners and the geometry with the bends shown.

To add the tabs, we need to create a new plane. Since the tabs are to allow the part to mount flush to another surface, the new plane needs to correspond to the front edge of the part.

FIGURE 12.65

Select Insert: Reference Geometry: Plane from the main menu or the Features group of the CommandManager. If there are any items already selected, clear them by right-clicking in any white space and selecting Clear Selections.

Select the edge shown in Figure 12.65, and then one of the front corner points, as shown in Figure 12.66. Click the check mark to create the plane.

FIGURE 12.66

FIGURE 12.67

View the plane from the right view to make sure that it is in the correct location, as shown in **Figure 12.67**.

Select the new plane. Sketch and dimension the first tab, near the upper-left corner of the part, as shown in Figure 12.68. Make sure to add a line along the bottom of the tab so that a closed contour is created.

FIGURE 12.68

FIGURE 12.69

Add centerlines, snapping to midpoints on the edges. Use the Mirror Entities Tool from the Sketch group of the CommandManager to place the other three tabs, as shown in Figure 12.69. To use the Mirror Entities Tool, select all of the lines and arcs to be mirrored and the centerline that the entities are to be mirrored about, and then click the check mark.

Extrude the tabs toward the rear of the part. Check the "Link to thickness" box, shown in Figure 12.70, to set the thickness as the same value that was chosen when the part was defined as a sheet metal part. Click the check mark to complete the extrusion.

FIGURE 12.70

Note that the tabs have sharp corners rather than bends, as shown in **Figure 12.71**. This is because the tabs were added after the bends were added in the FeatureManager. Rather than define a new set of bends for the tabs, we can rearrange them in the FeatureManager. Plane1 must also be moved, since it must be defined before the tabs can be added.

FIGURE 12.71

Click with the left mouse button and hold on the new plane (Plane1) in the FeatureManager. Drag the cursor up until the arrow points between the Extrude-Thin1 and the Sheet-Metal1 definitions, as shown in Figure 12.72.

Release the mouse and Plane1 will now appear before Sheet Metal in the FeatureManager, as shown in Figure 12.73.

Now click and drag the tabs (Extrude1) to immediately after Plane1, as shown in Figure 12.74.

FIGURE 12.72

FIGURE 12.73

FIGURE 12.74

The tabs now appear with bends, as shown in **Figure 12.75**. The metal around the bends was cut per the Auto-Relief parameters set earlier. (The width of the cuts is 0.5 times the thickness, and the cuts extend 0.5 times the thickness beyond the end of the bends.)

The cutout in the part must be added in the flattened state, so that it can be cut as a circle.

Drag the Rollback Bar to just below Flatten-Bends in the FeatureManager, as shown in Figure 12.76. To display the bend edges, switch to the wireframe mode (with hidden edges visible). On the front surface sketch and dimension the 25-mm circle as shown in Figure 12.77, using centerlines to place the circle.

FIGURE 12.75

FIGURE 12.76

FIGURE 12.77

Cut the hole with a through-all extruded cut. Show the part in the bent shape by dragging the Rollback Bar to the end of the FeatureManager, as shown in Figure 12.78. The finished part appears in Figure 12.79.

FIGURE 12.78

FIGURE 12.79

Save the part as "Sheet Metal Part."

While drawings of sheet metal parts showing their final dimensions are often made, a manufacturing drawing will show the dimensions of the flat pattern and *bend notes* detailing how the pattern is bent into the final configuration. We will now make a manufacturing drawing of the part that we just modeled.

Open a new A-size drawing, with a title block if desired. The Sheet Metal Part should be highlighted in the Model View Manager. Click the Next arrow, and choose the Flat pattern as the view, as shown in Figure 12.80. Click in the drawing area to place the pattern, as shown in Figure 12.81.

FIGURE 12.80

FIGURE 12.81

DOWN 70° R .16 DOWN 70° R .16

UP 70° R .16

UP 70° R .16

DOWN 70° R .16 DOWN 70° R .16

*Introduction to Solid Modeling
Using SolidWorks 2009*

Select the Options Tool. Under Document Properties, select Units and set the unit system to MMGS, with one decimal place for length dimensions and no decimal places for angles. Select Sheet Metal, and set the bend note style to With Leader, as shown in Figure 12.82. Click OK. The notes are now shown with leaders pointing to the bend lines, as shown in Figure 12.83.

FIGURE 12.82

FIGURE 12.83

Each bend note shows the bend radius and angle of the bend, as well as the direction of the bend (up or down) relative to the plane of the flat pattern. Importing dimensions into the drawing will not work correctly for most dimensions since the part dimensions are relative to the bent configuration. Instead, dimensions are added manually.

Add dimensions and notes as shown in Figure 12.84. Add a detail view to show the dimensions of one of the tabs. Save the drawing file.

FIGURE 12.84

Scale = 1:2
All dimensions are mm and degrees
Thickness = 2

Introduction to Solid Modeling
Using SolidWorks 2009

Chapter 12
Sheet Metal Pattern

PROBLEMS

P12.1 Open the interim assembly of the mold base and card holder that was created in section 12.2 (Figure P12.1). Perform an interference detection on the assembly. How are the interferences found related to the shrinkage factor used to create the cavity?

FIGURE P12.1

P12.2 Create a two-piece mold for the flange part of Chapter 1, allowing a 2% shrink factor. Figure P12.2A shows the lower mold half. Note that the material forming the holes is contained in the lower mold half. Figure P12.2B shows the upper mold half. Note that the chamfer feature is included in the upper mold half.

Hint: The easiest way to cut away material from the mold halves is to use *revolved cuts*. Figure P12.2C shows the sketch defining the "cutting tool" to be revolved around the centerline to create the lower mold half.

FIGURE P12.2A

FIGURE P12.2B

FIGURE P12.2C

P12.3   Create an assembly of the two mold halves of P12.2 and the flange.
Show a section view of the assembly, with the mold halves transparent,
as shown in **Figure P12.3**.

FIGURE P12.3

P12.4   Create a model of the sheet metal part shown in **Figure P12.4**. The
metal thickness is 0.060 inches, and the bend radii are 0.125 inches. All
dimensions shown are in inches.

FIGURE P12.4A

FIGURE P12.4B

P12.5   Create a model of the sheet metal part shown in **Figure P12.5**. The metal thickness is 0.060 inches, and the bend radii are 0.125 inches. All dimensions shown are in inches.

FIGURE P12.5A   **FIGURE P12.5B**

P12.6   Create the box shown from the parts created in exercises P12.4 and P12.5. Since the sides of the box are mirror images of each other, it will be necessary to have two different part files. Open the part created in P12.5, and select one of the vertical faces. Then select Insert: Mirror Part from the main menu. Click the check mark, and a new part file is created, linked to the original file. Save this file with a new name, and then assemble the three parts to form the box.

FIGURE P12.6

P12.7   Design a sheet metal bracket similar to the one shown in **Figure P12.7A**. The bracket is intended to be used to join a wood 2 × 4 (actual dimensions 1.5 by 3.5 inches) to a 4 × 4 (3.5 by 3.5 inches) post, as shown in **Figure P12.7B**, with nails or screws. Make a dimensioned drawing of the flat pattern of the part, including the bend notes.

FIGURE P12.7A

FIGURE P12.7B

CHAPTER 13

The Use of SolidWorks to Accelerate the Product Development Cycle

Introduction

Any company that designs and/or manufactures products has seen an increase in global competition over the past decades. To remain competitive, there is great pressure to develop new products. As a result, product models have shorter lives, and development costs are spread over a fewer number of units.

With these increased pressures, there has been a greater emphasis on improving and accelerating the product development process. While different industries and companies have their own unique procedures, there are some activities that are typically a part of a modern product development cycle:

- Physical prototypes are used early and often.
- Computer analysis is used extensively to complement physical testing.
- Engineering functions are performed simultaneously as much as possible, necessitating better teamwork and data sharing.

Solid modeling is an important tool in the product development process. The solid model becomes the common database used for a variety of engineering functions. In this chapter, we will explore two of the most common uses of solid modeling in product design: rapid prototyping and finite element analysis. We will also consider the challenge of managing and controlling the data produced during the product development cycle.

Chapter Objectives

In this chapter, you will:

- be introduced to the most popular rapid prototyping processes,

- see how the stereolithography file is used to define a solid part,

- learn how to create a stereolithography file,

- be introduced to the capabilities and limitations of finite element analysis, and

- learn about Product Data Management software.

13.1 Rapid Prototyping

The introduction of solid modeling made possible a new industry called Rapid Prototyping (RP), Additive Manufacturing (AM), Solid Freeform Fabrication (SFF), or Additive Fabrication (AF). Rapid prototyping refers to the creation of physical models directly from an electronic part file using an *additive* process. That is, the model is created by building it layer-by-layer rather than starting with a block and removing material to create the finished shape.

FIGURE 13.1

FIGURE 13.2

The first commercial RP method to be introduced was stereolithography (SLA) in 1987. In this process, a *photopolymer*, a liquid resin that cures under the application of certain wavelengths of light, is cured layer-by-layer by a precisely guided laser beam. The SLA process is illustrated in **Figure 13.1**. The part is submerged in a vat of photopolymer, supported on an elevator platform. The fabrication of each layer begins with the elevator lowering by the thickness of a layer, typically about 0.005 inches. The part is then covered by uncured resin. A mirror system directs the laser beam across the area to be cured for that particular cross-section. Laser power and the speed of the beam across the resin surface are controlled to ensure a complete cure of the resin while maximizing the speed of the process. When all of the layers are completed, which is typically several hours after the beginning of the build, the elevator lifts the essentially finished part out of the resin. Cleaning the part with a solvent and curing in a UV oven to complete the cure of the photopolymer are usually the only steps necessary to finish the part.

In the early 1990s, many new RP processes were commercialized. Among the most important was Selective Laser Sintering (SLS). In this process, plastic powder is *sintered*, or fused together, by the heat from a precisely guided laser beam. The SLS process is illustrated in **Figure 13.2**. As each layer is completed, the build platform drops by the thickness of one layer, and a fresh layer of powder is spread over the part and leveled. The laser is then directed onto the surface of the new powder layer, sintering the powder in the desired areas. The advantage of this process is that many engineering plastics, including nylon and polycarbonate, can be used. This process can also be used to make metal and elastomeric parts. Another advantage of the SLS process is that for complex shapes, the loose powder supports the part as it is being built. In most other RP processes, supporting structures must be built for complex parts, and then removed in a post-processing operation.

Another process that achieved commercial success was Fused Deposition Modeling (FDM), a process illustrated in **Figure 13.3**. The FDM process can be thought of as precisely directing a fine hot-glue gun to build up a part. The thermoplastic material (usually ABS plastic) is fed from a spool. There are actually two feed spools and extruder heads, as the material for building support structures is different from the part material. The support material is formulated to break away from the part easily, or to dissolve in water. The introduction of FDM was especially significant in that no laser or hazardous materials were required, allowing the machine to be placed in an office environment.

FIGURE 13.3

When RP was introduced, the available modeling materials had low strength and were extremely brittle. As a result, the only practical use of RP was to produce visual aids. This use was important, since the effective communication of complex designs can minimize errors in the interpretation. However, improvements in materials have allowed RP models to be used for many other functions. The Wohlers Report[1], an annual survey of the RP industry, found the uses of RP models in 2008 to be widely varied, and included:

Functional models	19.1%
Visual aids	15.3%
Direct digital/rapid manufacturing	14.9%
Presentation models	13.7%
Patterns for prototype tooling	12.5%
Fit and assembly	10.0%
Patterns for metal castings	6.5%
Tooling components	3.4%
Other	4.6%

The fact that an RP model can now be used in a functional prototype product or in tooling (which usually requires the longest lead time of all development activities) has increased the usefulness of RP greatly over the past few years.

Also, the growth of direct digital manufacturing (or rapid manufacturing) is an exciting trend. Rather than a prototype part, the final part is produced with direct digital manufacturing. This application of RP technology has the potential to make many low-volume parts cost effective, since no tooling costs need to be amortized. Also, parts that are difficult or impossible to make with conventional processes can be made with RP technologies.

---

[1]*Wohlers Report 2008, Rapid Prototyping, Tooling & Manufacturing State of the Industry Annual Worldwide Progress Report* (Fort Collins, CO: Terry T. Wohlers, Wohlers Associates, Inc.; wohlersassociates.com).

One of the biggest obstacles to increased use of RP has been cost. The systems described above range in cost from about $100,000 to well over $500,000. When the costs of maintenance, materials, and a trained operator are added, the costs of producing even a small RP model can be in the hundreds of dollars. Also, even *rapid* prototyping systems can be too slow at times. An engineer who has an idea for a new design concept in the afternoon and a design meeting the next morning probably will not be able to have a physical model in time for the meeting.

A new generation of RP machines has focused on lower costs and higher speeds, with some sacrifices made to accuracy and durability. The combination of low-cost materials and inexpensive off-the-shelf printing technology results in a fast and affordable (these machines cost about $30,000 or less) prototyping process.

FIGURE 13.4

These low-cost RP machines are called concept modelers or 3-D printers. More than 75% of the concept modelers sold in 2006 were based on one of two processes.[2] One of these uses the FDM process described above. The other is a powder-based process similar to SLS, but instead of using a laser to sinter the powder together, an inkjet printer head is used to spread binder fluid onto the surface of the part. A powder-based concept modeler is shown in **Figure 13.4**. Notice the cross-section of the ribbed flange part from Chapter 4 being printed.

Most domestic RP machines accept as input a type of file called a stereolithography (.stl) file. The structure of an .stl file is quite simple: the surfaces of a solid model are broken into a series of triangles, the simplest planar area. Each triangle is defined by four parameters: the coordinates of each of the three corners, and a *normal vector* that points away from the part, identifying which of the two faces of the triangle represents the outer surface of the part.

To illustrate how an .stl file defines a solid part, consider the simple part shown in **Figure 13.5**. Since this part has only flat, rectangular surfaces, two triangles can exactly define each surface, as shown in **Figure 13.6**. Since the part has six surfaces, the part can be described by 12 triangles.

FIGURE 13.5

FIGURE 13.6

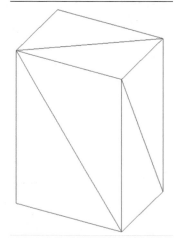

---

[2]Wohlers Associates Industry Briefing, May 2007, www.wohlersassociates.com.

To illustrate how an .stl file is created for a part with curved surfaces, consider the ribbed flange part modeled in Chapter 4, which is shown in **Figure 13.7**.

To create an .stl file from a SolidWorks part, select File: Save As from the main menu and select STL as the type of file from the pull-down menu, as shown in **Figure 13.8**.

FIGURE 13.7

FIGURE 13.8

Before saving the file, select the Options button. This opens the Export Options dialog box, and allows you to adjust the resolution of the .stl file, as shown in **Figure 13.9**. If the Preview option is selected, then the triangles of the .stl file are shown displayed in the part window.

FIGURE 13.9

The triangles created for a resolution set to Coarse are shown in **Figure 13.10**. Note the rough approximation of the fillets by the triangles.

The number of triangles created and the file size are displayed in the Export dialog box. For the Coarse resolution, 2326 triangles are used, and the size of the .stl file is 99 kb.

If the resolution is changed to Fine, then many more triangles are added to the curved surfaces, resulting in a smoother approximation of the actual shape, as shown in **Figure 13.11**. In this example, 6358 triangles are created, resulting in a file size of 318 kb. This file size is small enough to be easily transferred by e-mail or temporary storage media.

FIGURE 13.10   **FIGURE 13.11**

Usually, the Fine quality is sufficient. For parts where greater resolution is desired, the Custom quality can be selected, and the slider bars moved toward the right. The only trade-off for this higher quality is larger file sizes. However, since the transfer of large files via e-mail or Internet has become easier in recent years, file size is usually not a major problem.

When the .stl file is received by the computer controlling the RP machine, it is "sliced" into cross-sections by machine-specific software. Some machines allow for the thickness of the sections to be adjusted. Thinner layers result in greater accuracy, but of course longer build times. Most machines allow for multiple parts to be made simultaneously.

A model of the ribbed flange being removed from a powder-based concept modeler is shown in **Figure 13.12**. As with the SLS process, the loose powder around the model supports it during the build process. When the model is removed, it has low strength and must be handled carefully. The model is then infused with wax, cyanoacrylate adhesive ("super glue"), or epoxy to give it strength. The finished model, infused with epoxy resin, is shown in **Figure 13.13**.

FIGURE 13.12

FIGURE 13.13

13.2 Finite Element Analysis

Finite Element Analysis (FEA) is a method of predicting the response of a structure to loads by breaking the structure down into small pieces (*elements*). The locations where the elements come together are called *nodes*. Equations predicting the response of each element are then assembled into a series of simultaneous equations. Solution of this system of equations yields the displacements of the nodes. From these displacements, the *stresses* (forces per unit area) are calculated for each element. From the stresses, the factor of safety against failure of the structure can be predicted. The finite element method can also be applied to fluid flow analysis, heat transfer, and many other applications.

FIGURE 13.14

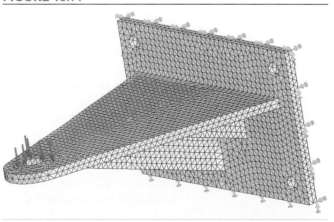

Solid modeling has enabled the increased usage of FEA, because it automates the task that in the past required the most time: creating the geometric model and the finite element mesh. A finite element mesh can be created from a solid model with only a few mouse clicks. **Figure 13.14** shows a mesh created from the bracket model of Chapter 3.

FIGURE 13.15

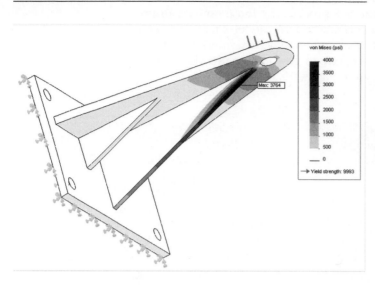

A 50-pound load will be applied to the end of the flange. After the mesh is created, the loads and *boundary conditions* are applied. Boundary conditions are the displacements controlled by external influences. For example, if we assume that the bolts attaching the bracket to the wall cause the back face of the bracket to be perfectly fixed, then we would apply corresponding constraints to the movements of the nodes on that face. The stresses resulting from the 50-pound load are shown as stress contours in **Figure 13.15**.

As FEA has become easier to use, it has also become easier to *misuse*. While the software takes care of the number-crunching, a knowledgeable engineer is required to set up the analysis and interpret the results.

Among the common errors made by inexperienced users of FEA are:

- *Inappropriate choice of elements.* Solid modeling software with a built-in FEA mesh generator allows a mesh of solid elements to be easily created from a solid model. However, in many cases solid elements are not the best choice. Relatively thin structures subjected to bending loads are usually better approximated with shell elements. A frame made up of welded structural members may require tens of thousands of solid elements to produce an acceptable grid, while a few beam elements will produce better results with substantially less calculation time.

- *Use of only linear behavior.* Linear analysis is based on the assumption that structural response varies linearly with loading. For example, the deflections and stresses of the bracket when loaded to 100 pounds are exactly double those produced by a 50-pound load. There are two types of nonlinearities possible, however. Geometric nonlinearity is present when the structure's stiffness is significantly different in the displaced configuration than it is in the original configuration. Material nonlinearity is present when the material is stressed beyond its yield point. When the material has yielded, it may not break, but will deflect further with little or no additional loading. Both of these nonlinearities require iterative solutions. That is, rather than applying all of the load at once, an increment of the load is applied and the equations are solved. The equations are then reformulated based on the results of the first solution, and solved again. This process is repeated until the solution is found. While this procedure requires much more processing time, it is necessary to obtain an accurate solution for many problems.

- *Inappropriate boundary conditions.* In the example of the analysis of the bracket, it was assumed that the back face of the bracket was fixed to a rigid wall. Is this a reasonable assumption? Maybe, but if the screws holding the bracket stretch slightly, then the top of the bracket will separate from the wall, placing more force on the bottom of the bracket. Most actual structural restraints are neither perfectly fixed nor perfectly free, requiring the engineer to use judgment in specifying the conditions to be placed on the model. In some cases, the analysis may be performed several times with different boundary conditions to obtain the limits of possible structural response.

- *Misinterpretation of results.* In **Figure 13.15**, the stresses displayed are the von Mises stresses. These stresses are calculated based on a specific failure criterion (the von Mises criterion) that is an excellent predictor of the yielding of ductile materials, but it may be inappropriate for predicting the failure of many materials.

These common errors are not presented to discourage the use of FEA, but rather to encourage the proper use of the method. An engineer should have a good understanding of mechanics of materials and at least an introduction to FEA theory before using FEA for any important application.

13.3 Product Data Management

We have mentioned earlier that solid modeling supports concurrent engineering, since many engineering functions can work from the same database (the solid model). Allowing multiple users to work on the same model or drawing, however, creates challenges, as well. How do companies manage their part files and drawings to control who can make changes?

Before answering this question, it is helpful to consider how companies managed paper drawings and other data before solid modeling was introduced. When a draftsman completed a drawing, it was checked and then went through a *release* process in which it was approved and signed off by various department representatives (design engineering, manufacturing engineering, safety, etc.). Copies were made and distributed, and then the released drawing was stored in a vault. Most medium-sized and large companies had a *configuration management* department that controlled the release process and subsequent changes. When a change was requested, a formal engineering change order process was followed. When the change was approved, it was amended to the drawing, or the drawing was modified and given a revision number, if the change was significant. It should be noted that many drawings would eventually have dozens of associated change orders, so managing this process was an important job. When a user of the drawing, say a stress analyst about to begin an FEA model, needed a copy, the configuration management department provided the latest version, including all change orders. Similarly, the Bill of Materials for each part was maintained and provided to the purchasing and manufacturing departments. In addition to controlling the drawing changes, the configuration management department would maintain *drawing trees* that showed how drawings related to each other.

A change in a part could affect an assembly at the next level. Other data might have been controlled but not necessarily linked to the drawings. For example, a stress analysis report might have been released and stored in the vault, but a person looking at the drawing would probably not know that the analysis had been done, or where to find it. Similarly, in industries where weight is important, mass property reports were produced, but usually not linked to drawings.

The introduction of solid modeling presented many new challenges to configuration management. Some of the features that make solid modeling so exciting to design engineers—associativity between parts, assemblies, and drawings, ability to use the solid model for many functions—could cause huge problems from a data management perspective. That is, one person could make a seemingly simple change to a component and that change would propagate throughout an entire assembly without the person being aware of its effect. Now that some companies model major systems with solid modeling, controlling who can make changes is vitally important.

Product Data Management (PDM) software is used to perform some of the tasks of the configuration management department, while streamlining communications between various departments. PDM entails two broad categories of functions:

- Data management, the control of documents (part files, drawings, stress reports, etc.), and
- Process management, the control of the way in which people create and modify the documents.

The data management function is similar to that of the paper-based configuration management department, except that the released drawings are now electronic files and are stored in a *virtual* vault instead of a physical vault. (Actually, backup tapes and disks of the virtual vault are often stored in a fire-proof physical vault.) Since the part and drawing information is stored in a relational database, immediate location and retrieval of files is possible. This helps to reduce redundancy, especially among standard components such as rings and fasteners. A design engineer who needs to specify a fastener can easily determine if there is already a similar fastener with a part number assigned in the system. If there is, then the purchasing department will not have an additional item to buy.

Process management is the control of active procedures: who generates the data and how the data are transferred from one group to another. One important feature of PDM systems is work history management. In a paper-based system, having old, outdated drawings around is an invitation for trouble, since they can be used by mistake. But by destroying old drawings, the history of the modifications made to that drawing can be easily lost as well. PDM software can track the change history of a part, an important function in Total Quality Management (TQM), while protecting against accidental usage.

What does the future hold? Most manufacturing companies implemented Manufacturing Resource Planning (MRP) systems long before PDM became popular. MRP systems allow the tracking of raw materials, work in progress, and finished inventory in the plant. The move to reduce inventories and adopt just-in-time raw materials deliveries necessitated the adoption of MRP systems. In many ways, implementing PDM to engineering functions is analogous to implementing MRP for manufacturing.

The logical next step, then, is to combine MRP and PDM systems into a single system referred to as Product Lifecycle Management (PLM) or Enterprise Resource Planning (ERP). As with any new large-scale system, implementation is expensive and time-consuming, and so these systems are not widely used in most industries.

A recent example of the complexity of engineering data and the challenges of managing that data is the delay announced in 2006 of the production of the Airbus 380 jumbo jet. Both Airbus and its competitor in the commercial aircraft market, Boeing, have suppliers around the world. Coordinating the work of these suppliers is a monumental challenge. On October 3, 2006, Airbus CEO Christian Streiff announced that delivery of the 550-seat A380 would be significantly delayed because of data translation problems between engineers in Germany, who were designing and building the wiring harness for the plane, and engineers in France, where the final assembly of the plane was taking place. As a result of the errors, the wiring harnesses would not fit correctly—a major problem for a plane with hundreds of miles of wiring. The problems were expected to cost Airbus over $6 billion dollars in profits.[3]

13.4 Some Final Thoughts

The widespread adoption of solid modeling has been part of a revolution in the way that products are designed and developed. As many companies have thrived with new technology, others that have not kept up have not been able to compete and have been forced to close. Although no one can predict the future, one thing seems certain: the pressure on companies to develop new products faster and better will not lessen.

On a more personal scale, the same concept has held true for many engineers. An engineer who is adverse to change and reluctant to learn new tools is at a great competitive disadvantage to his or her peers.

The good news is that most people who enter the engineering profession do so because they have the curiosity to want to learn new things. Keeping that curiosity alive will allow an engineer to have a rewarding career that is always fresh and interesting.

---

[3]"PLM: Boeing's Dream, Airbus' Nightmare," Mel Duvall and Doug Bartholomew, *Baseline*, February 1, 2007, Vol. I number 69, Ziff Davis Media.

INDEX

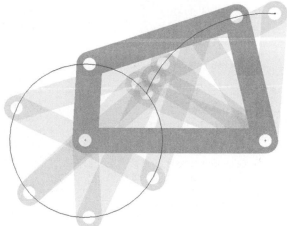